D0323337

Dark Ages

Dark Ages

The Case for a Science of Human Behavior

Lee McIntyre

A Bradford Book
The MIT Press
Cambridge, Massachusetts
London, England

MIT Press books may be purchased at special quantity discounts for business or sales promotional use. For information, please e-mail special_sales@mitpress.mit.edu or write to Special Sales Department, The MIT Press, 55 Hayward Street, Cambridge, MA 02142.

This book was set in Stone sans and Stone serif by SNP Best-set Typesetter Ltd., Hong Kong and printed and bound in the United States of America.

Library of Congress Cataloging-in-Publication Data

McIntyre, Lee C.
Dark ages : the case for a science of human behavior / Lee McIntyre.
p. cm.
"A Bradford book"
Includes bibliographical references and index.
ISBN-13: 978-0-262-13469-9 (alk. paper)
ISBN-10: 0-262-13469-1 (alk. paper)
1. Social sciences. I. Title.
H85.M45 2006
300.1—dc22

2006046173

10 9 8 7 6 5 4 3 2 1

For Josephine
Per Ardua

Men at some time are masters of their fates:
The fault, dear Brutus, is not in our stars,
But in ourselves, that we are underlings.
—William Shakespeare, *Julius Caesar*, Act I, Scene II

Contents

Acknowledgments

I have benefited greatly from the advice and suggestions of several friends and colleagues who were kind enough to provide comments on earlier drafts of this book: Rich Adelstein, Jon Haber, Harold Kincaid, Noretta Koertge, Dan Little, Mike Martin, Dan McIntyre, Susan McIntyre, Andy Norman, Alex Rosenberg, Merrilee Salmon, and Pat Starr. To each I offer my heartfelt thanks, especially given my certainty that none of them will agree with all that I have written here.

I would like to give special thanks to Kit Ward, who believed in this book almost before anyone else did, and to my editor Tom Stone, who believed in it when it most counted. I would also like to thank my copy editor Beverly Miller for her keen eye and good suggestions, which saved me from a number of infelicities, as well as Susan Clark, Robyn Day, Sandra Minkkinen, and all of the others at The MIT Press who have helped this book to see the light of day. I am especially grateful to my friend Laurie Prendergast for preparing the index.

Finally, I thank my wife, Josephine, who has always believed in me and supported my ideas. It is the happiness that I have felt in our life together that has inspired me to think about how to make the world a better place and has given me the confidence to write this book.

Introduction

A few years ago, in the small town in which I lived in upstate New York, there was a villagewide power blackout. Although it was nearly 2:30 in the morning, I couldn't resist going outside. Nothing was lit; the night was as black as the ocean. Looking up, I could see the thousands of stars that are normally lost to the light pollution of modern civilization. As I did so, I began to reflect on what it must have been like to live in the Dark Ages of human life, before the scientific revolution and modern technology. Of course, the Dark Ages weren't literally dark; presumably they had sunlight during the day and torches and candles at night. But the metaphor is apt; for total darkness does make one reflect on the progress that humans have made and the scope of what we today take for granted.

What would it feel like to live in a Dark Age? Would you realize it? Or would you just see the achievements of the day—perhaps even feeling lucky to live in such "modern times"—and fail to see all that had *not* been achieved. Of course, no one living in a Dark Age would call it that; rather this label is placed on a backward era only by a later one, in which the state of human civilization is more advanced. With the benefit of hindsight, it is easier to see what has been missed. But isn't there nonetheless some way to judge one's own era?

Look around you. We live in a time of enormous technological achievement, when we are able to bend nature to our will, and yet we suffer from the same social problems that have plagued the human race for millennia. Despite the enormous progress that we have made in our understanding of nature, who can honestly say that the bulk of the problems that are the cause of human misery today are not of our own creation? And yet what have we done about them?

The comparison between our success in understanding nature and our failure to understand ourselves is vast. We have satellites and fax machines that transmit stories of barbarous cruelty that could have been told by our ancestors. We have ever more sophisticated weaponry of war and yet no true understanding of what causes war in the first place. Terrorism, crime, war, and poverty continue unchecked throughout the world, largely because we lack the understanding to stop them. We are as ignorant of the cause-and-effect relations behind our own behavior as those who lived in the eighth or ninth centuries once were of those behind disease, famine, eclipses, and natural disasters. We live today in what will someday come to be thought of as the Dark Ages of human thought about social problems.

What were the first Dark Ages like? And why are they called that? The Dark Ages are one of the most intriguing periods of human history. They mark a nearly 600-year blank spot in the progress of human civilization in which the knowledge of antiquity almost completely disappeared from the West. It was a time when few people received any sort of education whatsoever, and life was governed by the superstitions and fears fueled by ignorance. In terms of the exploration of ideas and the quality of human life, the Dark Ages were indeed dark; they always seem to me a temporal analogy to the huge blank spaces on ancient maps of the world that are marked "unknown."

Although scholars differ as to the exact beginning and ending dates of what should properly be called the Dark Ages, they are normally taken to occupy the period of time from the fifth to the eleventh centuries A.D., sometimes also known as the Early Middle Ages. The Middle Ages themselves occupy a nearly 1,000-year span of time between the period of classical antiquity (which reached its height in the Greek and Roman empires and ended with the fall of Rome in 476 A.D.) and the Renaissance of the fifteenth and sixteenth centuries in Europe (which saw the rebirth of learning in the West). Hence the name "Middle Ages" is given to the time between these two great eras in which human knowledge flourished, which lasted from roughly the 400s to the 1400s A.D.

What is the difference between the Dark Ages and the Middle Ages? Scholars of the medieval period will be quick to point out the political and economic changes that occured during the High Middle Ages of the twelfth century and the important ideas of various Scholastic thinkers from this period. It is a mistake, they will argue, to use the terms *Dark Ages* and *Middle Ages* synonymously. And yet—in terms of the creative advancement of human thought beyond the dominant paradigm of medieval Christianity—there were no significant breakthroughs in art, science, philosophy, or literature during this time.

Then came the Renaissance, first in Italy, then to spread throughout Europe. During this period, thinkers began to recover, and to be influenced by, the learning of antiquity, and great advances were made in art and literature. The philosophy of humanism was born, and with it came a focus on the role that human reason might play in directing the course of our lives, challenging the hegemony of Christianity.

The Renaissance came last to the sciences, beginning with Copernicus's publication of a new theory of the universe in 1543, which spawned the scientific revolution of the sixteenth and seventeenth centuries that brought us Kepler, Galileo, and Newton. During this period, scientists sought and discovered many of the great laws of nature by employing a methodology using experimentation, the application of mathematics, and a belief that there was a natural order to the universe. It is the success of this viewpoint that has led to all of the modern achievements of science, even to the present day. Indeed, if one is bold enough to defy historical convention for a moment in order to focus exclusively on the sciences, one might usefully—albeit loosely—think of the Dark Ages for the natural sciences as extending from the period following the birth of scientific reasoning by the Greeks and the great technological advances of the Roman Empire (which ended in the fifth century), all the way to the scientific revolution of the sixteenth and seventeenth centuries—a period of almost 1,100 years.

Seen from this historical context, it is clear that scientific reasoning is a rare and fragile thing and that its advances can be impeded or even extinguished altogether if given the wrong set of cultural conditions. Indeed, knowing this, we would be wise to ask ourselves what remaining barriers might stand in the way of the extension of scientific reasoning into new domains. Did the scientific revolution extend as far as it might have? Can scientific inquiry be brought to the remaining areas of human ignorance, far beyond those probed even by the greatest minds of antiquity and the Renaissance? Now that we have seen how successful science can be in answering some of the long-standing questions that humans have asked about the natural world, might we employ it in trying to understand the greatest remaining blank spot on the map of human knowledge: the causes of human behavior?

Of course, the skeptics will say that this is impossible—that despite the best efforts and ambitions of the Enlightenment, many thinkers have tried and failed to establish the scientific study of human behavior. From Francis Bacon to the logical positivists, the holy grail of understanding human action within the context of a precise scientific calculus has remained an elusive dream of Western thought. But why? Many will argue that it is because human beings are unique and not subject to scientific treatment. Are we not conscious of the forces that move us? Do we not have free will to change them?

In the years that have passed since the inception of the scientific revolution, numerous scholars have meditated on the reasons that we have waited so long for a revolution in the social sciences. Grown bitter over time, the consensus of such scholars today seems to be that such a revolution is never coming. In the interim, an impressive array of arguments has been formulated that attempt to show why it is impossible to employ a scientific mode of understanding in the study of human behavior. And, like jilted lovers, the number of scholars who are receptive to such arguments has grown over time, until today the voices against the prospects for a science of human behavior drown out all others.

In advocating a science of human behavior, I have taken seriously the idea that there are powerful arguments against it, and I have spent the better part of my career examining them. A few years ago, I wrote a scholarly book, *Laws and Explanation in the Social Sciences: Defending a Science of Human Behavior*, in which

I systematically analyzed all of the major arguments against a science of human action. And the conclusions I came to were shocking. Not only were the arguments weak, they were so weak that I became convinced that even their advocates did not really believe them. I began to see that by and large, the arguments put forward were not meant to convince someone who seriously wanted to advocate a science of human behavior. Rather, they were mere window dressing—a peg on which to hang the prejudices and fears of those who did not want to have a social science in the first place. Thus, I came to the conclusion that the primary reason that we do not today have a science of human behavior is not that it is impossible; it is that we fear the threat that such a science might pose for our cherished religious or political beliefs about human autonomy, environmental determinism, race, class, and gender. In short, I believe that—just as in the study of nature 400 years ago—the primary barrier to a science of human behavior is ideological. Political ideology is today doing to social science what religious ideology did to natural science in the first Dark Ages.

Some things never change. Resistance to knowledge has always characterized periods of great scientific advance. When a new paradigm threatens the reigning religious or political order, we manufacture congenial but weak arguments against it. Many of the contemporary arguments against a science of human behavior are rooted in a naive misunderstanding of the nature of scientific progress and work against a true "social science" only if we are prepared to believe that science has already had its last victory. Time and again, however, science has overcome such prejudices and replaced convenient myths with testable theories. The scientific truths that we today take for granted seem obvious to us only because of the courage of those who fought for them against earlier prejudice. I argue that we must now be willing to make this same effort on behalf of the new scientific frontier: empirical inquiry into the causes of human action.

I came to write this book precisely because I think that the majority of philosophers and social scientists working today do not have such courage, and have sought to smother the public's desire for a precise understanding of our social problems under the forces of political correctness. Afraid of what we might find out about ourselves, today's academics have stood in the way of a science of human behavior in precisely the same way that religious clerics attempted to stunt the scientific revolution of Copernicus and Galileo. Having abdicated their responsibility to improve the human condition, many of today's scholars satisfy themselves with the status quo in social science, feeling that it is preferable to preserve the myths that we harbor about the causes of our actions rather than attempt a systematic study that may topple the idols of political fashion. Thus does ideology take precedence over empirical investigation; we fiddle while the world burns.

By contrast, I hope to show that there is something that we can do about the current situation in the social sciences—that just as the human race once saw its way clear from the ignorance and superstition that had dominated its thinking about nature, culminating in the scientific revolution, we may now take the first steps toward a social scientific revolution in which we come to understand the true forces behind our social ills, and so may build an improved human society on this basis. But, I argue, we may do this only if we take seriously the idea that we have a long way to go in our understanding of human behavior and that the only way to get there is to follow the path lit by science.

This is not a traditional academic book. I owe its inspiration, however, to two scholars, both now dead, who were not content to see their work have influence only in academic circles, but instead sought to bring learned debate to the attention of a larger audience, in the hope that by addressing some of the great social dilemmas of our time, we could do something about them. The first is someone I never met, James Harvey Robinson, who believed in putting learning in the service of human betterment and also in the hands of the public. His important book *The Mind in the Making* (1921) (a best seller in its day) long ago highlighted the folly of the human condition: When we have the tools to improve our situation, why do we tarry?

The second scholar is someone whose life touched my own and who served as an inspiration of a first-rate scholar who never lost sight of her obligation to improve the social world. Barbara Wootton's *Testament for Social Science* (1950), like Robinson's earlier book, sought to make the case for a science of human action, at a time (just following World War II) when we most needed to hear the message.

In this book I have sought to follow in their footsteps and engage a wider audience in what is arguably the most important debate there is over the future of the human race. The stakes could not be higher. The issues at hand affect us all and should not be locked up in the hands of only a few professors. With this in mind, I have tried to write this book in an accessible style, with few footnotes and virtually no professional jargon, in an attempt to reach the broadest possible audience. In doing so, I hope to engage those readers, both inside and outside the academy, who long to make social change by improving the horrifying social conditions that will long affect human life until we have the courage to do something about them.

1 Diagnosing the Human Condition

It was the best of times, it was the worst of times, it was the age of wisdom, it was the age of foolishness, it was the epoch of belief, it was the epoch of incredulity, it was the season of Light, it was the season of Darkness.
—Charles Dickens, *A Tale of Two Cities*

As we begin the twenty-first century, how many of us can honestly say that we are optimistic about the direction in which the human race is headed? Aren't the social problems that we face today much the same as—or worse than—those our ancestors faced? Yet aren't they for the most part of our own making? But what have we done about them?

Ask yourself what it is today that contributes most to human misery and suffering. In the first Dark Ages, the answer was easy: disease, hunger, and natural disasters. Today we have modern medicine, grow more than enough food to feed the planet (though we still have famine, largely for political reasons), and understand the causes of earthquakes, volcanoes, and hurricanes well enough that it is possible to ameliorate their effects on us through advanced warning.

In modern times, however, the list of ills that we face seems quite different. The tyranny that we face today is largely a result of human cruelty and indifference in the way that we treat one another: women cannot walk the streets alone at night for fear of sexual assault; inner-city children cannot use public parks for fear of death from a stray bullet; child prostitution is rampant in Southeast Asia; modern slavery exists in the Sudan and Mauritania; Congo's civil war so far has resulted in 3 million deaths in five years, including beheadings and other atrocities; systematic rape was used as a form of terrorism during wartime in Bosnia and has since been emulated in Congo and the Sudan; genocide in Rwanda resulted in 800,000 deaths during three months in 1994; Sudan's civil war has caused 2 million deaths and created 4.5 million refugees; 6 million Jews were killed during the Holocaust; countless Korean "comfort women" were forced into sexual slavery by Japanese soldiers during World War II; during the "Rape of Nanking," soldiers held contests to see how many people they could kill in an hour; school and workplace shootings have become commonplace; prison rape is routine; serial killings are on the rise; and on a crisp September morning a few years ago, the world witnessed a new form of human horror when the World Trade Center in New York City was demolished by terrorists.

How can we be so cruel to one another? Why are such things allowed to happen? At times it seems that there is nothing that is so horrible that it has probably not been done by one human being to another at some time on this planet. But the worst thing to realize is that such atrocities continue largely because we tolerate them or because those of us who will not tolerate them do not know where to turn for answers. Can anything be done to stop such modern evils? What are the causal forces behind them?

One might imagine that such questions are the proper domain of the social sciences. First conceived of in the nineteenth century as the scientific study of human action—on analogy with the natural sciences as the study of nature—the first social scientists self-consciously emulated the natural scientific model of explanation and sought to do for our understanding of human behavior what Newton had done for the study of nature. In realizing this goal, however, the social sciences have been a dismal failure.

In testament to the success of Newtonian physics, humans first began to assume their mastery over nature, leading to the technological breakthroughs of the industrial revolution, culminating in our own day in space travel and digital computers. In the application of our understanding of human social behavior to the amelioration of human misery, however, the social sciences have few such successes to offer. For all of the theoretical and statistical apparatus of criminology, do we really feel that we know how to reduce crime? For the much-vaunted technical precision of economics, do we have confidence that we know how to avoid a recession? For all of the studies that political scientists have done on terrorism, has agreement been reached on the question of whether concession to terrorists incites further demands? Do the experts in these areas even have solid theories on which to base good social policy? On September 11, 2001, where were the social scientists—the experts in psychology and sociology—who might have helped us to understand the larger forces at work? Are we convinced that we now understand the root causes of terrorism so that we may prevent such actions in the future?

In its current incarnation in the contemporary disciplines of economics, psychology, sociology, anthropology, political science, and history, much of today's social scientific inquiry does not even seek to address the pressing social problems of our day, choosing instead to focus on technical or interpretive theoretical models of human action that make such grand simplifying assumptions that they have little predictive or explanatory value when applied to the real world. Faced with this, many of the practitioners of social science have begun to deny that their disciplines should even be conceived of as scientific. Yet the scope of human suffering continues unabated.

We all read the same newspapers and watch the same news on TV, tallying the scope of human misery as a result of war, crime, and poverty. And yet, despite all of our scholarship, what have we done to remedy these problems? Aren't the evils we suffer at our own hands pretty much the same as those we have been inflicting on one another through the ages? Will we ever be able to make any progress? Should we simply give up?

Ignorance and indifference are the two biggest obstacles to changing our fate. I am an optimist. I believe that the majority of people are not cruel and do not wish to see others suffer; that is, I believe that most of us would not be so indifferent or resigned to our inhumanity to one another if we were not so ignorant of how to stop it. But we *are* all currently ignorant of the true nature of the causal forces behind riots, depressions, strikes, and wars. Is this because there are not cause-and-effect relations behind them? I know very few people who actually believe this. Rather, it is that we do not understand the nature of science and how it can be applied to the study of human affairs. And (I reveal a little pessimism here) I believe that we are also afraid of what we may find out about ourselves if we look too closely.

The fact that we are ignorant of the causal factors behind our social ills is overwhelmingly apparent. To demonstrate this, let's examine something that almost everyone would agree is an important social problem that we would like to prevent: major crimes such as murder.

On May 6, 1996, a story appeared on the front page of the *New York Times* with the headline: "Major crimes fell in '95, early data by FBI indicate." Good news. Perhaps if we understood why the drop in the national crime rate had occurred, we could fine-tune our social policy and decrease the crime rate for years to come. The text of the article following the headline, however, gave us no such hope. When asked to explain the drop in crime rate here is what three leading experts in criminology had to say:

"I think we can now say a trend has been established. . . . But I'm not sure we know why."

"When push comes to shove, nobody has any ability to explain the increases any better than the decreases. . . . Criminologists are like weathermen without a satellite. We can only tell you about yesterday's crime rates."

"The honest answer is that no one knows why crime rates go up and down."

What possible reason might there be for such utter helplessness? Criminology, after all, might seem to be one of the best hopes for a true social science. We have been collecting data for centuries, we have the ability to study not only group but also individual behavior in a relatively controlled setting so that we may test hypotheses (e.g., does parole contribute to recidivism?), and we have statistical models to help us put the demographics into perspective. So why the absolute failure to come up with testable theories?

One might think, on the basis of the quotations just given, that criminologists could think of no theories whatsoever, or that they were unable to test their assumptions against the data.

Is this the case? Some possible theories: the decline in crime was related to an improving economy in the United States in the 1990s, the decline in crime was a function of tighter immigration enforcement, the decline in crime was a function of fewer criminals being on the street due to increased law enforcement efforts and longer prison sentences, the decline in crime was merely an artifact of a demographic shift whereby fewer people were in their crime-committing years. These are all testable hypotheses. Indeed, the last one mentioned, noted by a fourth criminologist in the *New York Times* article, is quite interesting in that it implies that the drop in crime rate could have been predicted. Thus, as demographic patterns shift again in the next decade, as a new and larger generation of those in their crime-committing years comes of age, we can expect the major crime rate to increase once again.

It seems entirely appropriate that criminologists examine such hypotheses and that they ask the difficult questions that they raise about the relationship between crime and economic status, race, IQ, gender, and age. I am not a criminologist. And I am sure that all of the above hypotheses have occurred to those who are. So what is it that is lacking in our understanding of crime? Why do we remain so ignorant? Is it for want of better data? For lack of understanding the scientific tools available to us? Or for want of the intellectual courage to test our favored hypotheses? Are we willing to have a science of criminology? Or shall we forever remain ignorant of one of the most pressing sources of human misery?

Lest the reader think that this is only an isolated example, and that surely we are better off in our other attempts to understand the causes behind our social afflictions, let's examine two more cases.

The first was the momentous occasion in 1991 when the Soviet Union fell, ringing in a new era of political and social order for the largest country on our planet. To be sure, there were many uncertainties facing the changeover from a planned socialist economy to a fledgling capitalist one. Unanticipated results, like the explosion in organized crime, made for a bumpy road to a free market economy. Perhaps the most formidable and threatening barrier to a capital economy, however, was hyperinflation, which was fully predictable by social scientists! That is not to say that the social scientists had done their job in anticipating hyperinflation in post-Soviet Russia, any more than a meteorologist has done his job in telling us that we are in for some snow in upstate New York in February. We want to know why hyperinflation would inevitably follow from the destruction of the socialist economy and, more important, what we might do to stop it.

For those who have not lived through hyperinflation, it is easy to underestimate its human consequences, and scoff at its comparison to a problem such as crime. But it causes genuine misery. During 1993, Russia had an inflation rate of 300 percent per month. Stop for a moment and imagine this. You get a paycheck in July, and by the time you get your next one, its value has been reduced by two-thirds. The upward spiral in prices, and wages trying to catch up to them, heads up and up until you get to the point where Boris Yeltsin announced in mid-1996 that he was going to multiply war veterans' pensions by a factor of one hundred as a gesture of goodwill. The breathtaking thing, however, is to realize that had he wanted to really make them keep pace with currency inflation over the previous five years, he would have had to multiply them by a factor of ten thousand!

What can be done about this terrible social problem? Would limited price controls help? Should the market be allowed to continue as it is until it reaches equilibrium? Will it ever reach equilibrium? Should the money supply be restrained? Where were the answers given by social scientists?

It is surely obvious to anyone who has studied economics that economists do have an answer to the cause of hyperinflation. Hyperinflation—defined as the sharp and sudden fall in the value of currency and a rise in prices—is caused by an increase in the supply of money and the psychological factors attendant on a loss of confidence in the currency. Moreover, as opposed to the creeping inflation that we all are fairly familiar with, hyperinflation is normally associated with some enormous social upheaval such as occurred in postwar Germany in the 1920s and Hungary in the aftermath of World War II.

What we really want to know from the social sciences, though, is whether hyperinflation can be stopped altogether. That is, rather than simply knowing the factors associated with hyperinflation, we want to know enough about it to prevent its dire consequences when we see it coming. It seems simple enough: If we genuinely understand the causes of hyperinflation, why are we not able to control it given sufficient control over the money supply or the ability to impose price controls? Do we really understand all of the causal factors at work? Have we gotten to the bottom of the problem? Is our understanding sufficient to improve the quality of our lives when faced with such a problem? Such questions go to the heart of social scientific inquiry into factors like hyperinflation, where we require knowledge sufficient not merely to explain—or even to be able to predict—such eventualities, but to be able to control their impact on our lives.

The second case is about one of the worst breakdowns in social order in the past two decades. It seems beyond human comprehension to imagine how any society, even one pushed to the brink during wartime, could invent something so horrible as the "ethnic cleansing" used by the Bosnian Serbs against the Muslims during the war of the 1990s. Not only were thousands of Muslims killed, but soon the Serbs developed a new and chilling twist on ethnic cleansing, involving the systematic rape of Muslim women by Serbian soldiers. Indeed, the systematization of the crime was central to its commission. This was not some random wartime atrocity committed by a few morally heinous individuals. It was, instead, a program by which 20,000 to 50,000 Muslim women were taken captive and raped until they became impregnated by Serbian men.

The moral outrage of most Americans was matched only by the utter ineptitude of our political leaders to do anything about it. Short of war, we were told, there was no instrument available to us to stop this atrocity. And leaders' response, that atrocities were being committed on all sides, seemed to say that to know that *others* were suffering too somehow made it more palatable. As a nation, our post–World War II commitment "never again" to allow the systematic and state sponsored victimization of one ethnic group by another that occurred during the Holocaust became a hollow promise, as ethnic cleansing occurred while the world watched.

It is hard to know what to say about such an utter breakdown in our humanity to one another—not just by the perpetrators but also by the spectators. It is easy to say that it was the American political leaders who failed to do anything. Or that the American people did not care enough about the problem to pressure their government into action. When faced with worldwide inaction in the face of a contemporary Holocaust, there is more than enough blame to go around. Where it leaves many of us, however, is searching for answers to the question of how this could happen at all, let alone twice (or more) in one century. What must possess a mind to commit such horrible crimes against another person? What is the human response to authority when given orders that are patently immoral and illegal? Can the entire episode be chalked up to diffusion of responsibility or the attractiveness of defending morally heinous actions with the claim that one was "just following orders"? Or are there other—perhaps unknown—forces at work that motivate and captivate the human mind toward cruelty? Can we commit right now "never again" to allow something like this to happen because we will immediately begin trying to explain its causal roots in the human psyche?

Such, as I see it, is the charge to social science: to study the most pressing social problems of our day so that we may understand the causal forces behind them, and therefore hope to control them better. It is not enough to comprehend roughly or vaguely the tendencies in such social events. *There are genuine and specific causal forces behind them.* And until we understand the nature of these causal factors, we will continue to be at the mercy of our ignorance and suffer its devastating human consequences.

It is, of course, probably unknowable whether, even if we had a better understanding of the criminal mind, or hyperinflation, or wartime psychology, any of the particular instances I have cited here could have been averted. But as long as our tools in fighting against them are so blunt that we often feel utterly helpless to do anything at all, certainly more understanding would be an improvement. And that is what I am asking of the social sciences: to take seriously the idea that there are real cause-and-effect relationships behind the social problems of our day and to believe that these can be discovered through the proper scientific inquiry. Indeed, let us indulge ourselves for a moment and try to imagine what it may be like in the far-off day when social science hits its mark. To be sure, there will still be social problems. For all of the success of natural science and modern technology, we still have not conquered AIDS or cancer, or entirely figured out how to shelter ourselves from the caprice of natural disasters. But the issue I am after here is that in natural science, we have hope; we feel that we are on the right track. We feel that if such questions are answerable, they will be addressed as a result of scientific inquiry. What, then, of the prospects for success in social science?

Imagine a society where we understood individual psychology well enough to know the predispositions for criminal behavior or pathological obedience. Imagine a world in which we had studied war as a scientific problem, not in the terms put forth by "military science" (which is the study of military tactics and strategy), but rather in order to help us to understand how war might be averted, in much the same way as game theorists have studied human tendencies in cooperation and retaliation. Imagine a social order in which we were not subject to sudden economic upheaval because we understood well enough the relationship among inflation, unemployment, and the supply of money.

Of course, many will dismiss these imaginings as nothing more than a vision of a social utopia that is no more reachable to us than the flights of fancy found in science fiction. But let's think about that for a moment. To a mind of the first Dark Ages, what would our contemporary technological society have seemed like if not a utopia? We have electricity and automobiles, we have airplanes and inoculations against childhood diseases, we have satellites, computers, fax machines, microwaves, VCRs, photocopiers, the Internet, and all of the other technology that we today take for granted. Such is the bounty of our conquest of the laws of nature that resulted from the scientific revolution. Is such a conquest of the social order now available to us as a result of a genuine science of human behavior?

Such a conception, of course, is not new. It is as old as the Enlightenment, when the fledgling social sciences first looked with envy at the achievements of the natural sciences. And yet the history of such attempts to embark on a true scientific study of human behavior has not lived up to its promise. But why? Is it because such a program is impossible in principle and could never succeed no matter what our efforts? Is human behavior just not amenable to study by scientific methods? Or is it that this way is open to us, and perhaps we are already on it, but the progress is just much slower than we would like? Or, perhaps, are the barriers at work of a different kind altogether, resulting largely from the fact that we do not want to have a scientific understanding of human behavior because we are afraid of what we may find?

As I shall argue in this book, the improvement of the human condition awaits our response to such questions. What will we do to answer them? The fruits of a greater understanding of our social problems are, I hope, obvious. And I believe that a better grasp of the causal forces behind these problems is essential to our hope to improve the quality of human life. The path toward this, I shall argue, is lit by science. It holds great promise in helping us to understand the laws of human nature in just the same way as it has been successful in helping us to discover the laws of nature. A genuinely informed and enlightened social policy might then be able to come forward.

Yet, as we shall see, there are powerful forces at work to block this path. And so the world burns as we linger yet in these Dark Ages.

2 A Science of Human Behavior

What it demands is that the sociologist put himself in the same state of mind as the physicist, chemist, or physiologist when he probes into a still unexplored region of the scientific domain. When he penetrates the social world, he must be aware that he is penetrating the unknown; he must feel himself in the presence of facts whose laws are as unsuspected as were those of life before the era of biology; he must be prepared for discoveries which will surprise and disturb him.

—Emile Durkheim

In chapter 1, I recounted some of the horrors with which the social world presents us and our ignorance of the causal factors behind them. I argued that the only way to improve the human condition is to have a better understanding of these causal forces, and I suggested that science is well equipped to help us to do this. In this chapter I will continue to advance the idea that the solution to our social ills is a genuine science of human behavior. Indeed, it is our only hope. If compassion alone would work, it would have; if relying on our intuitions about human nature would work, it would have; if trusting in our leaders' wisdom would work, it would have. But nothing has worked.

Yet is this surprising? Our efforts to the present day seem more like trial and error than they do systematic inquiry. This is because none of the attempted solutions to our social problems so far have been based on a sustained and objective attempt to understand the causal forces behind our action. Our behavior does not happen at random; there are reasons behind each of the social ills detailed in the previous chapter, just as there are reasons behind the laws of nature. But so far we have made no systematic study of what these causes are and have been trying to design our social policy based largely on our interests and prejudices and suppositions. The result has been our utter ignorance of the causal factors behind our social problems and our consequent failure to do anything about them.

The best way to overcome such ignorance is science. Indeed, if there are real causal forces behind our action and they are empirically discoverable, what better method might we have than science to unlock their secrets? In the past, science has worked where nothing else has. It is our most powerful method of gaining knowledge. Thus, it would seem reasonable to assume that science is uniquely capable of helping us to gain a better understanding of our actions. But just as our behavior does not happen by accident, neither does science. We must make a concerted effort to cultivate a science of human behavior before we can hope to have one. And before we may emulate the standards set out by science, we must be sure that we are not misunderstanding what science is most basically about.

Let me deal at the outset with an obvious objection. Hasn't this already been tried? Do we not currently have hundreds of thousands of social scientists who are studying these matters? If a true science of human action were possible, wouldn't we have found it (or don't we have it) already? Yes, I concede that we do have fields that are called "the social sciences," but as I shall argue, these inquiries for the most part have not been very scientific. By saying this, I do not mean to disparage the few valiant social scientists who have made a real effort to do good empirical work and who are as disgusted with the present state of our ignorance of social affairs as I am. Indeed, part of the problem is not only that we have not had very much scientific inquiry into the causes of human behavior, but that—to the extent that there has been good empirical work—it has been widely ignored and underpublicized. This is surely due, at least in part, to the phenomenon whereby bad social science drives out good; if policymakers and others feel free to lump all social science together as ideologically driven, and thus dismiss it as they would any other political view with which they don't agree, they will miss much good empirical work. Still, there are some examples of good social science, and I will point these out along the way throughout this book.

Rather, what I am suggesting is that the lion's share of what today passes for social science is ridiculously unrigorous and bungles even the good pieces of data that are available to us so far. And whether they emulate or eschew the standards set out by science, it is my contention that most social scientists working today have either misunderstood or ignored the true nature of scientific inquiry. Thus, I believe that today's social scientists fall into one or more of several camps: (1) they want to have a genuine social science but do not know how to go about it; (2) they are trying to have a genuine social science but are just having a difficult time overcoming the barriers to empirical social research; (3) they believe that scientific inquiry into human affairs is not appropriate or possible; (4) they practice a kind of inquiry in which the theorizing and analysis is so infected with ideology, prejudice, and wishful thinking that whether they emulate scientific standards or not, they will fail to achieve useful results no matter how hard they try.

To those in the first category, the remedy is obvious: social scientists should undertake a close analysis of scientific methodology before they set out to examine social affairs. They must learn about the dangers of naive correlation, manufacturing untestable hypotheses, and all of the other methodological faults that may flaw scientific research. For those in the second category, the solution is to study not just scientific methodology but also the history of science in order to see how the natural sciences overcame barriers that once seemed insurmountable. It is easy to overestimate the barriers facing social science if one does not realize that they were once (and are still) faced by natural science. To those in the third category, I ask them to read on and tell us which one of the ridiculously weak arguments against the possibility of a science of human behavior they favor. Moreover, I ask them to tell us which of the several alternatives to scientific inquiry has a better chance of improving the human condition. Finally, I ask them to be honest about their motives and to be sure that they are not just making an excuse for believing something that is congenial to their emotions. To those in the fourth category, which I think accounts for the bulk of social scientists, the solution is much more difficult, for what is at stake here is as much a function of having the right attitude toward scientific work as it is in knowing what procedures to follow.

Genuine science is a rigorous and systematic method of inquiry by which one compares a theory against empirical evidence that may potentially prove that theory wrong, and a willingness then to abandon or modify the theory, where necessary, in light of the evidence gathered. The most important thing about science, therefore, is the role of evidence in testing and shaping our scientific theories. It is not enough for our theories to be based on empirical observations or for them to fit some of the data of our experience. *They must be put to the test.* We must attempt to compare our theories not merely against our intuitions or look for the instances in which they fit our observations. We must try to find out where they are weak and where they do not fit our observations; we must gather new evidence and see how our theories hold up. We must, in short, take a theory that we may hope very much to be right and try our best to prove that it is wrong.[1]

Ironically, one of the best demonstrations of scientists' reliance on the power of evidence can be seen in what some commentators have called the worst example of scientific bungling in the twentieth century: the search for "cold fusion." On March 23, 1989, two chemists, B. Stanley Pons and Martin Fleischmann, held a press conference at the University of Utah to announce that they had achieved a sustained nuclear fusion reaction at room temperature. If true, this result would have vast scientific and technological (and financial) implications, since it would mean that someday soon the world would have access to a clean, cheap, and virtually inexhaustible supply of energy. Although many were enormously skeptical of the result—due in part to the fact that Pons and Fleischmann were announcing their results somewhat prematurely in a press conference rather than the more customary route of publication after peer review—the scientific world set about at once to test the experiment.

In short order, it was revealed that Pons and Fleischmann's results were unreproducible and that their procedures had been methodologically flawed. Rancorous debate ensued, and accusations of extrascientific meddling ran wide. Yet in the end, the appeal to empirical evidence was all that mattered. Although many scientists were profoundly embarrassed by the whole episode, and many of the books that have been written about it have had titles such as *Bad Science* or *Cold Fusion: The Scientific Fiasco of the Century*, one might argue that rather than revealing science at its worst, the cold fusion episode was science at its finest! Indeed, in testament to the power of science, despite all of the politics, money, and prestige involved, in a relatively short time the dispute had been settled based on only the appeal to empirical evidence. Although one particular theory had seen its demise, a victory had been won for the method of science.

As we can see from this example, what is distinctive about science as a way of knowing is that it has a built-in check against our hopes and intuitions; it is a method by which we constantly make sure that we aren't just fooling ourselves in what we believe to be the case, by comparing our beliefs against the data of experience. One realizes, therefore, that apart from any list of ideal methodological criteria, what is crucial in having a genuine science is the cultivation of the scientific attitude toward evidence: A propensity to doubt that we already basically know the right answers to empirical questions before we have engaged in empirical inquiry. A willingness to test our cherished beliefs and to be surprised by what we may find out. A willingness to change our beliefs on the basis of what we have learned as a result of our scientific investigations. This is what is essential in fostering the scientific attitude. And this is what I claim is missing in most of the current practice of social science.

Most modern social science falls far short of this ideal. Routinely, social scientific inquiry fails to tests its hypotheses against empirical evidence even where it is possible, infects its theories and testing procedures with ideological assumptions, and conducts social inquiry within a framework of unrealistic and untestable hypotheses about human nature that insulates them from ever learning the secrets behind our behavior. Worst of all, it seems that the mass of contemporary social scientific inquiry is constrained by the curious conviction that social scientific theories must be intuitively plausible or must fit neatly within certain ideological parameters. What scientific progress might be made by further empirical observation is then stalled through a reluctance to abandon old congenial theories, or to embrace or even fully examine new and controversial ones.

To say that the social sciences are not now scientific, however, does not amount to saying that their theories are not based on any evidence; there is obviously some attempt to reconcile theory with data. Rather, it is to say that what is missing from social inquiry is a healthy respect for the true power of testability—the willingness not just to explain our theories in light of the evidence, but to abandon or modify them because they do not fit the evidence.

Often in social scientific practice, even where evidence is used, it is not used in the correct way for adequate scientific testing. In much of social science, evidence is used only to affirm a particular theory—to search for the positive instances that uphold it. But these are easy to find and lead to the familiar dilemma in the social sciences where we have two conflicting theories, each of which can claim positive empirical evidence in its support but which come to opposite conclusions. How should we decide between them? Here the scientific use of evidence may help. For what is distinctive about science is the search for negative instances—the search for ways to falsify a theory, rather than to confirm it. The real power of scientific testability is negative, not positive. Testing allows us not merely to confirm our theories but to weed out those that do not fit the evidence.[2]

But this is precisely what has not occurred in the social sciences. Social science is choking on a plethora of bad theories because we have not allowed the evidence alone to decide between good and bad theories and have failed to discard even those theories that have long since been refuted. This can be seen most starkly in social science textbooks, where theories that are hundreds or even thousands of years old seem to be recycled ad infinitum, despite mountains of disconfirming evidence. Aristotle's physics has long since been discarded, but his politics is still taught by social scientists as if it reflected fresh insight into the human condition.[3] Is this due to the depth of his insight in this one particular area or, more likely, that we have progressed so little in the political realm that we have nothing better to offer? Can we do no better in social science?

A good example of this failure to come to grips with the negative power of evidence in social scientific research can be seen in the ongoing contemporary debate over the relationship between immigration and welfare. Are immigrants more likely than natives to be on welfare? Do immigrants receive more in services than they pay in taxes? Clearly this topic of research, more than most others in social science, has the potential for impact on public policy. Predictably, it is also therefore one of the most highly politically charged debates in all of social science. Sadly, despite the apparent simplicity of engaging in straightforward empirical investigation to answer these questions, the politicizing of this dispute has prevented the emergence of a consensus answer.

Early studies on the relationship between immigration and welfare seem obsessed with providing support for the ideological view that immigration is not a net drain on the economy, and indeed that it is "good for America." This is shown by attempting to prove that immigrants "pay their own way." Not surprisingly, however, there has been little agreement among social scientists: some studies have shown that immigrants clearly pay their own way, others have shown that they clearly do not, and still others have claimed that immigrants confer a net economic *benefit* to the American economy.[4] What might account for such a disparity? Isn't the answer to this question one that can and should be decided on the basis of empirical evidence?

Unfortunately, each side of the debate has accused the other of sullying the data with ideological bias. The "conservative" side has accused the "liberal" side of cherry-picking statistics that artificially lower the numbers on welfare participation by immigrant groups, by conveniently ignoring "age" and "cohort" effects. In turn, the "liberal" side has accused the "conservative" side of similarly using faulty estimates on several parameters, resulting in an overestimation of service costs due to immigrants and an underestimation of taxes collected. As if this were not bad enough, a few liberal critics have at times resorted to ad hominem claims that the bulk of the work that purports to show that immigrants do not pay their own way has been done by one researcher of Cuban ancestry, George Borjas, who is himself an immigrant but who now wants to "pull up the ladder" behind him. Magnanimously, Borjas has emphasized the empirical nature of the issue at hand, even while admitting that this debate is "fraught with questionable assumptions, which effectively determine the answer to the question."

In other words, in this debate, as in so many others in the social sciences, we can always find evidence to confirm whatever hypothesis we like, so long as we are prepared to embrace a given set of ideological and methodological assumptions that support it. But this is to infect the method of social research with ideological considerations. It therefore slights the empirical possibilities of social scientific research, whereby we hope to learn from our evidence before we decide on a policy, rather than the other way around. This strategy flies in the face of the scientific attitude of dispassionate inquiry, guided not by our desire for any given outcome but instead by what can be learned from our data. At best, this could not help but tempt some researchers to ignore and deemphasize evidence that did not support their hypotheses. At worst, it could lead to outright suppression of contradictory data. Either way, this would seem to subvert the whole commitment to letting empirical disputes be decided on the basis of the evidence by letting us get away with the assumption that we already basically know the answer to most social scientific questions on the basis of our ideologies or our intuitions. Are we not here very far away from the standards that were used ultimately to decide the dispute over cold fusion?

As the example above shows, what is needed in social science is agreement that empirical standards alone are the appropriate means for adjudicating social scientific disputes, so that we may weed out outdated theories and advance those that are more accurate. Of course, it could be objected here that in social science, it is an ideological issue even to decide what should count as evidence. Yet in this, the social sciences are no different from the natural sciences, where the scientific attitude is put under similar pressure due to the subjective bias that some researchers may have for particular theories. But what allows natural science to go forward even in the face of this is the universal assent had by natural scientists that such extraempirical attachments are base, and should be eschewed, as they are likely to lead to problems such as those witnessed in the initial dispute over cold fusion. And besides, they are bound to be uprooted by others, who are trying to prove our theory wrong.

Thus, it is not simply that contemporary social science does not use evidence, but that the way that it uses evidence is too unrigorous, and too infected with ideology, to amount to a truly scientific understanding of human behavior. In social science, in contrast to natural science, it seems that by the time one goes in search of empirical evidence, a favored theory has already been chosen, and evidence is being gathered not in order to test it but in order to confirm it. In gathering evidence in the social sciences, we often come to a point where we have a choice to make: What parameters, what assumptions, what exclusions will guide our choice of what counts as evidence? There is a great temptation to let this be decided by ideology, for if we already know what theory we would like to support, it is not difficult subtly to cull the evidence in its favor.

Contrast this with the use of evidence in the natural sciences, where it is expected that researchers will bend over backward not to make assumptions that are favorable to their theories and where there are elaborate methodological safeguards (like double-blind clinical trials) in place to guard against any residual bias. Indeed, any bias in natural scientific theories is potentially short-lived, as there are normally dozens of other researchers in place who are eager to try to reproduce new findings, who would soon expose any bias.

In natural science, we take the refutatory power of evidence seriously and seek to test our theories—to prove them wrong—

long before anyone else sees them. Natural science is a competitive enterprise, where careers are made not only by discovering new theories but also by refuting those that have been taken for granted. In social science, however, we often seek to gather evidence in order to prove our theories right, and hope that no one questions or even knows about the assumptions that we have used to shape our inquiry. Is it any surprise that the results of such inquiry have been less than scientific? We may be using evidence in such social hypothesizing, but we have ignored the empirical power of evidence to help us weed out false theories. Yet if we are genuinely interested in learning from our data, in pursuit of a scientific understanding of human action, shouldn't we agree that the same standard of evidence that is employed in the study of the natural world must be embraced in the study of human behavior?

This change in attitude alone, I contend, would be the beginning of a revolution in the social sciences. Based on the proper understanding and use of the scientific attitude toward evidence, who knows how far the social sciences may reach? When we begin to treat our own behavior as a proper subject for scientific investigation—instead of blithely assuming that we already know all there is to know about it—who can foretell what we may learn about the causes of our behavior? Thus, the failure of social science today should not in any way be taken to indict the possibility of a genuine social science in the future, any more so than the 500-year-long failure of medieval alchemy should be held against modern chemistry. If we are not conducting our inquiry in the proper way, is it any wonder that success eludes us? Shall we set about at once to reform the social sciences, on the belief that social science, like natural science, should be driven by taking a scientific attitude toward evidence?

Now, I know what the response to all of this will be. I can already hear the cries! "But in the social sciences, we face a set of barriers unlike those in the natural sciences. Our subject matter, after all, is ourselves, and there are unique problems that result from this fact." Thus arise several of the most commonly cited objections to the prospects for a science of human behavior.

Common Objections to a Science of Human Behavior

1. *The subject matter of social science is prohibitively complex.* This claim has many different variations, all of which come down to the same basic idea. Science has studied comparatively simple systems. But the subject matter of the social sciences is human social relations, which are far more complex than the subject matter of any of the natural sciences. The number of variables at work in influencing any individual's behavior is so large that it is all but impossible to study them scientifically. By the time we have a theory that might account for the causal influences behind even one piece of human behavior, things have already changed.
2. *Human behavior is part of an open system.* The fundamental assumption behind this objection is that science is possible only when we are studying a closed system, with a finite number of variables. In the social sciences, however, it is claimed that the number of possible influences on human behavior is potentially infinite. If it is predicted that I will go downtown for lunch as usual, this may be subverted by any number of unforeseen factors: I may fall down and skin my knee, I may hear on the radio that the bridge is closed, or I may decide that I just do not want to go today. One cannot hope to have a science that would control for all of these unforeseen factors.

3. *It is impossible to be objective about our own behavior.* Unlike in the natural sciences, this objection tells us, we care deeply about what our social inquiry may reveal about human behavior. We are indifferent to the laws of nature. We are not so indifferent to the causal forces that may motivate our actions. Given the fact that every social scientist is also a human being, we are confronted with an inherent subjectivity in social science that does not exist in other kinds of research.
4. *In social science we cannot perform controlled experiments.* To a large extent, social inquiry is historical. But historical events cannot be repeated to see how they might come out if some one factor had been different. We cannot rerun an election to see how it might have come out without a third candidate. Moreover, even in studying individual human behavior, we face a historical element, for in social science, we are dealing with a subject matter that learns from each experimental trial. Thus, we cannot perform a controlled experiment because we cannot ask the subject to forget what he or she has learned from the first time we ran the experiment.
5. *Humans have free will.* At base, the most fundamental objection to the possibility of a science of human behavior is that as human beings, we are free to act in any way that we choose at any time that we choose. Once we become aware of any purported regularity in our behavior that may be discovered by social science, we are free to change it.

Such objections are standard and are commonly cited as fundamental barriers to the possibility of a science of human action. But as I shall argue, they are *all* misfounded or overestimated and fail to show that we could not have a true social science. In an earlier book, *Laws and Explanation in the Social Sciences: Defending a Science of Human Behavior,* I examined in detail each one of these, and many other, objections raised by philosophers, social scientists, and others, against the prospects for a science of human behavior. And what I found was shocking. Not only did each of the arguments decidedly fail to show that we could not have a genuine social science, they failed so miserably that I became convinced that they could not even be the real reason that most people are against a science of human action. Given their weakness, how could they be? Instead, I came to the conclusion that such arguments are merely window dressing, serving as convenient cover for most people's strong emotional desire not to believe in a science of human behavior.

The real reason for most people's objections—their fear that the scientific examination of our behavior would threaten our deep-seated ideological prejudices about human nature—I shall discuss in chapter 3. These objections too fail, but for different reasons, which I shall identify. For the remainder of this chapter, however, I will confine my remarks to the demonstration that each of the five common arguments against a science of human behavior fails. (For more argumentative detail than I am able to provide here, readers should refer to my earlier book.) What I hope to establish here is that but for our prejudices and our attraction to the weak arguments in their service, a genuine social science would be possible.

Refutation of Common Objections

1. *The Argument from Complexity.* The flaw with the argument from complexity is not that it identifies some feature that is not present in social scientific inquiry, but that it overestimates the extent to which complexity stands in the way of science. First, it is important to be clear on one feature of the argument that

is often ignored: What is "complexity"? Surely complexity is not some absolute metaphysical property, which is a function of the way that things are in the world "as such." Rather, complexity is a function of our descriptions of the world. Thus, it is misleading to say that the subject matter of social science (or any science) is just complex, for this is to speak as if complexity were not relative to our theories and categorizations of the phenomena under scrutiny. But as anyone who has engaged in scientific inquiry knows, complexity can be ameliorated by progress in our understanding of the relationships at hand.

It is not that science studies simple systems. It is that science *simplifies* the subject matter under investigation by revealing the basic causal connections that are at work. Indeed, one problem with the argument from complexity is that it encourages us to underestimate the stunning complexity faced by natural scientific inquiry every day. Moreover, it correspondingly tempts us to overlook the very foundation for the success of science. The hallmark of science is its flexibility in developing new theories and descriptions of familiar phenomena so that we may understand the causal factors that are at work behind them. Relative simplicity in the subject matter under investigation, given the descriptions and theories we are using, may facilitate the search for credible theories. But science has an impressive track record in dealing precisely with those cases in which our subject matter does seem, at first glance, to be prohibitively complex and where we cannot find simple and straightforward relationships between an antecedently given set of variables. Indeed, at such times, science must reinterpret the phenomena under investigation before it can explain for us what they most basically are. In this way, science may simplify a previously complex subject matter by providing a more basic understanding of what it is that we are trying to explain. Science is perfectly well equipped to deal with complexity. Complexity therefore is not a fundamental barrier to social scientific inquiry.

2. *The Argument from Open Systems.* The problem with the argument from open systems is to be found in its reliance on two implicit assumptions: that human systems are open in the first place and that science cannot effectively be used in the study of open systems. The problem with the first assumption should be obvious. Although it may seem intuitively plausible to us that the number of possible influences governing our behavior is potentially infinite, what evidence do we have that this is really true? How do we know that the number of influences is not just incredibly large, in which case the real objection is one about complexity?

But even assuming that the number of potential influences over human action is infinite, what leads us to believe that scientific inquiry is therefore impossible? While it is true that the classical conception of science is Newtonian—where we study a closed system with a finite, and rather small, number of possible variables—contemporary science has provided numerous examples whereby science has made progress in studying open systems. Evolutionary ecology, meteorology, and in particular the new science of chaos provide examples where science has succeeded in the systematic study of systems in which the number of potential variables may well be infinite.

Science is about the systematic study of regularity. But who is to say that there is no regularity in open systems? A system that is open need not be lawless. Infinity is not the same as indeterminacy. The real barrier to science isn't infinity but the lack of order that might accompany it. But why suppose, especially in light of the great regularity in human behavioral patterns, that human systems are unordered? Science seems perfectly capable of studying open systems as long as they present some behavioral regularity. If there is causal regularity behind human action, we may study it with science. Thus, even if human systems are open, unless we are prepared to accept the idea that our behavior is random, we could still have a science of human behavior.

3. *The Argument from Subjectivity.* We now face one of the most popular, and misunderstood, objections to a science of human behavior: the allegedly ineliminable "subjectivity" of social inquiry. The problem with this view is not that subjectivity is not a real problem in social science. The problem is that subjectivity is often used as an excuse for bad social science. Yes, it is hard to be objective in the study of our own behavior. Indeed, it is now something of a truism in the philosophy of science that one cannot even be fully objective in the study of nature. For do we not there too bring our interests to bear on our scientific research? But in social science, as in natural science, this does not mean that we should not try to do the best job that we can. Indeed, it is worth pointing out here that if the philosophers are right, and we cannot achieve perfect objectivity in any scientific inquiry, then one may rightfully ask why the social sciences cannot hope to do at least as well on this score as the natural sciences? After all, it is these same flawed self-interested human beings who conduct natural science who do social science as well; the success in pursuing objectivity could not in social science, any more so than in natural science, depend on scientific practice only by perfect individuals. As we saw in the example on cold fusion, natural science too has individuals who prefer their own theories, are inclined to ignore evidence, and seek the spotlight. But science as a whole is more objective than its practitioners, for it progresses by discovering the mistakes of

others. Indeed, it is the appeal to evidential standards that keeps science as objective as it is. Thus, wishful thinking in science is rooted out not because of unique intellectual honesty among scientists, but because of the fear of public embarrassment against objective standards.

Yet here we face what may well be a real difference between the practice of natural and social science. In natural science, even if one realizes that it cannot be achieved perfectly in practice, there is a healthy respect for objectivity as an ideal. The extent to which we bring ideological factors to bear on the scientific investigation of nature is to be repudiated and is recognized by its practitioners as a danger to good scientific research. Yet in the social sciences, the importation of our political and social interests seems to be part and parcel of doing social science. In most social science, as opposed to natural science, we have comparatively little fear that our ideological assumptions will be exposed to refutation by hard facts. The infection of social scientific theories by our personal agendas does not seem to carry the shame that it does in natural science. The recognition of the value of empirical inquiry, unadulterated by our interests, seems missing in social science. Thus, in social science, as opposed to natural science, there seems to be little respect for objectivity even as an ideal.

It is perhaps true that objectivity is harder to come by in social science than it is in natural science. In social inquiry we are, after all, dealing with ourselves as the subject matter. But what is wrong with social science, I believe, is that the inability to have perfect objectivity is too often used as an excuse to abandon the objective ideal altogether. And this, in turn, leads to the failure to have a scientific attitude toward evidence in the social sciences.

This is not to say that in social science the evidence is ignored completely, or that social scientific theories are made up out of whole cloth, with no effort made to conduct experiments. It is that in social science, unlike in natural science, the practitioners are not nearly so embarrassed about mixing up their ideological assumptions with their empirical inquiry. Thus, in social science, more so than in natural science, the evaluation of evidence seems beholden to the interests of those who are conducting the inquiry. The result is that social scientific theories are so loaded with ideological bias from the very beginning that even where evidence is gathered, it cannot be used to jettison bad theories. This is because what is bad about the theories is their ideology, which is probably not subject to empirical verification in the first place. Thus, what passes for social science these days is often that someone who already knows what outcome he or she would like to have, and what policy he or she would like to advocate, gathers evidence to back up his or her theory. Thus does lack of respect for the standard of objectivity lead to repudiation of letting social scientific inquiry be driven by evidence. And this leads directly to what I have identified as the absence of a scientific attitude in the social sciences. Indeed, some social scientists have embraced this fact and have even used it to criticize those who have attempted to appeal to empirical evidence in support of their social scientific theories. In one such instance Charles Leslie has argued:

> Most of the influential work in the social sciences is ideological, and most of our criticisms of each other are ideologically grounded. Non social scientists generally recognize the fact that the social sciences are mostly ideological, and that they have produced in this century a very small amount of scientific knowledge compared to the great bulk of their publications. Our claim to being scientific is one of the main intellectual scandals of the academic world, though most of us live comfortably with our shame.[5]

But must it be like this? Does the lack of perfect objectivity in practice necessitate the abandonment of the objective ideal and a corresponding absence of the scientific attitude? Must the ineliminable subjectivity of our interest in ourselves color the way that we must perform social inquiry? Of course, we cannot turn off our interests, hopes, and fears, even when we are engaging in science. But what we can do is attempt to keep them from infecting our inquiry. Indeed, there is enormous value in doing so.

Just look at medicine. Here is an example of a science in which we have tremendous vested interests. And yet we realize that there is no purpose served in allowing wishful thinking to influence our analysis of how things are. Our overriding interests in medicine, as in social science, are normative. We have a practical goal in mind when conducting our inquiry, and we know at the outset what it is that we value. Yet in medicine we do not use normativity as an excuse to abandon the objective ideal. We recognize that normativity is different from subjectivity and indeed that our normative goals might be undermined if we were to give up our pursuit of objectivity. Should the situation be any different in social science?

Aiming at objectivity, even where it is known to be unreachable, is an important aspect of scientific inquiry. It means that you are allowing yourself to learn as much as you can from the evidence and that you are not intentionally biasing what your experiments are trying to tell you. It means that you have a scientific attitude toward your inquiry. Respect for objectivity, I claim, is essential for scientific inquiry. Subjectivity need not serve as a barrier to pursuing a science of human behavior, therefore, anymore than it does in natural science, if we resolve not to let it stand in the way of our cultivation of the scientific attitude.

4. *The Argument from Lack of Controlled Experiments.* The objection that we cannot have controlled experiments in social science is one of the weakest objections against the possibility of a science of human behavior. For why think that it is essential to science to have controlled experiments? This objection is usually favored by those who do not have a very sophisticated understanding of the natural sciences. For to those who have any familiarity with natural science, it is obvious that many sciences succeed without the luxury of controlled experiments. Astronomy, geology, meteorology, and evolutionary biology all deal with systems where the opportunity for controlled experiments, for one reason or another, is limited. And yet they are all scientific. This objection therefore fails to show that the lack of controlled experiments stands in the way of scientific inquiry and so that it would prevent a science of human behavior.

It is also important to point out that there are many more opportunities for experimentation in the social sciences than have been realized. Social psychology, in particular, makes extensive use of experimentation. And yet the results of the experiments have often not been accepted as scientific. Why not? Perhaps the problem is not with the experiments but with the low quality of the theories that are being tested. An experiment can help you to learn something only if you are sure what it is telling you and if you are willing to modify your theory accordingly. Methodological flaws in knowing how to learn from the tests that are performed on their theories therefore may be preventing social scientists from learning as much as they can from their experiments. Concentrated study of the dangers of naive correlation, unfalsifiable hypotheses, and unrealistic assumptions would do more to aid the scientific progress of social science than to pine for more opportunities for controlled experimentation.

5. *The Argument from Free Will.* We now face what is taken to be the most formidable objection to the prospects for a science of human behavior: that our behavior is fundamentally unpredictable because we have free will. I shall not pretend that this argument is an easy one to dismiss. For the most part, this is because of the powerful human intuition of freedom of the will, juxtaposed with the lack of empirical evidence to decide this question. We *feel* free and so we must *be* free. And for most people, that is the end of the story.

The implications of the argument from free will for a science of human behavior, however, are far from clear. On its surface, it may seem like a knock-down argument against the idea that we can discover the "laws of human nature." For if we are truly free, there are no laws governing us at all. We can act in any way that we see fit.

My response to this argument may be interpreted as a pragmatic vindication of a science of human behavior. It is based on the premise that we do not know whether we have free will and that we must decide whether we should try to have a science of human behavior in the absence of decisive evidence concerning this question. How then should we proceed? My claim is this: if we do have free will, fine. We still would need a social science, for it seems clear that even if we are free, we are as yet ignorant of the alternatives to the actions we perform that are making us so miserable. If we are free, then why do we keep making the same mistakes over and over? Why is there such stunning similarity over the centuries between the crimes that we commit against one another if we are free to stop them? Indeed, if we are free, isn't our behavior all the more horrible? *Especially* if we are free, we need a social science to help us realize it and liberate us from the habits that are the source of so much human suffering. If, on the other hand, we do *not* have free will, then social science would surely be necessary to help us learn our limits. It would help us to understand and accommodate to whatever forces there are that are controlling our behavior. There would be an end to certain social policies that were based on unrealistic views about our capabilities, to be sure. But more important, we would finally realize that there are limits to human malleability and may figure out how to have the best society that we can within our capabilities. If we may someday discover the blueprints of human nature, may we not then construct an improved social order to fit it? *But either way, we will need a science of human action.*

There is one final piece of business here that needs attending, for people often bring up the argument from free will as if they already knew which horn of the dilemma we are on: whether we actually have free will or not. But how could they know that? It is an empirical question whether we have free will. And the whole motivation for social science is that we do *not* already know everything that there is to know about ourselves merely by intuition. It is that we may learn something by making ourselves the subject of scientific study. What hubris to think that we can decide the ultimate question of our being—our very freedom—solely on the basis of intuition. It begs the question against the very project of social science to say that we know whether we have free will. Thus is social scientific inquiry vindicated no matter whether we have free will, for either way, we cannot be sure, and we must decide which path to pursue in our understanding of ourselves in the meantime.

We have now seen that all of the major objections to the possibility of a genuine science of human behavior fail. The claim that there is a fundamental difference between the subject of study in social science and the study of nature, thus necessitating a different method of investigation, is mistaken. Indeed, it is worth reflecting for a moment on why these arguments fail and what the implications are for the future of social science.

First, each of the arguments (with the reasonable exception of the argument from free will) fails because it does not identify factors unique to the social sciences. That is, the very same factors listed above—complexity, openness, lack of controlled experiments, and subjectivity—that allegedly militate against a science of human behavior, if they were decisive would also show that we probably could not have a science of nature either. For in natural science as in social science, we suffer from many of the same problems. Thus, what is revealed is the fact that by and large, the criticisms of the prospects for a science of human behavior are based on a naive and overly idealized view of the barriers facing the study of nature. Of course, if one has an overinflated view of the natural sciences, the prospects for a social science will look dim. But this is due not to any real disanalogy between natural and social scientific practice, but rather to the artificially high standards of most critics of the social sciences (many of whom have never systematically studied natural science in the first place).

Second, this must mean that the barriers just cited are perfectly appropriate ones to be faced (and overcome) by science. Science has demonstrated tremendous ingenuity in the face of methodological barriers like objections 1 through 4 listed above and should give us confidence that they may be overcome in social inquiry as well. The matters at issue in social science, no matter how subtle they may seem to us, are at base empirical, and therefore are best handled by science.

Indeed, third, note that the spurious barriers cited above could not be decisive in preventing all genuine social science, because there is some very good social science that faces these barriers every day. The empirically focused inquiry of evolutionary psychology—which is based on the idea that the theory of evolution by natural selection can explain not only human physiology but also human behavior—is a good example of the effort to apply scientific standards to the explanation of human action, even under the constraint of complexity. Likewise, Daniel Kahneman and Amos Tversky's work on judgment under uncertainty involves the study of human reasoning under carefully controlled experimental conditions, revealing several core tenets of rational judgement that human beings regularly violate. This work, in turn, has spawned the field of "behavioral economics," in which economists embrace experimental methods and eschew the once dominant "simplifying assumptions" that for so long have kept economic theory out of touch with economic reality. In the field of criminology, Gary Kleck has done some of the most rigorous empirical work ever seen in the field on the relationship between guns and violence, puncturing the mythical assumptions behind several deeply held beliefs about guns (which will be discussed in an extended example in chapter 4). Each of these is an example of social scientific inquiry that faces the very barriers cited above and whose success belies them. Critics of a science of human action would therefore do well to stop making excuses for why empirically focused social science cannot go forward; philosophers in particular should stop citing allegedly fundamental a priori barriers to the scientific study of human behavior that are daily being eroded by good social science.

Finally, recall my earlier claim that it is not surprising that the arguments analyzed in this chapter are so weak, since they are not the real reasons that people do not believe in a science of human behavior in the first place. Rather, they are a convenient set of excuses for those who do not wish to know things that might make them uncomfortable. Indeed, there is a strong undercurrent in the litany of reasons that we cannot have a genuine science of human behavior that we should be overjoyed that this is the case.

But why? Why are most people so horrified at the prospect of the scientific study of human behavior, and why are they so eager to embrace the weak arguments against it? What are the real reasons people do not want to have a genuine science of human behavior? And what has actually stood in the way of its realization lo these many centuries? That is the subject of my next chapter.

3 Resistance to Knowledge

The reason social scientists do not more often arrive at the truth is that they frequently do not want to.

—Bertrand Russell

If one may be allowed to hope where one does not know, then I hope from my heart that these investigators and microscopists of the soul may be fundamentally brave, proud, and magnanimous animals, who know how to keep their hearts as well as their sufferings in bounds and have trained themselves to sacrifice all desirability to truth, *every* truth, even plain, harsh, ugly, repellent, unchristian, immoral truth—for such truths do exist.

—Friedrich Nietzsche

My thesis in this chapter is as easy to state as it is hard to accept, and yet in our hearts each one of us knows it to be true. It is that we resist knowledge that is not congenial to us. What do I mean by this claim? Simply that we human beings are egotistical creatures, who enjoy indulging in the fantasy that our opinions actually influence how things are in the world. We believe that if we deny the evidence or refuse to study certain taboo topics, then the facts about them are not real for us. That is, we often avoid gathering knowledge that is likely to clash with our favored theories and high opinion of ourselves.

Do you doubt it? Do you think this is a general phenomenon that doesn't apply to you? For readers who remain unconvinced, try the following test to see if you have opinions on topics about which we as yet have no definite empirical evidence or, worse, where you find yourself ready to doubt well-confirmed theories simply because you hope that they aren't true. Do you approach empirical matters dispassionately? Or do you already know how you would like things to turn out? Do you have intuitions about empirical matters before data have been gathered? Do you have a vested interest in certain theories turning out to be correct?

Ask yourself this: Why is it that some of those reading this book right now are hoping that I am wrong and that we cannot have a science of human behavior? What is it that is so appealing about being unpredictable? What is so threatening in admitting that our behavior is a potential object of scientific study, and that perhaps we do not know all of the reasons behind our actions? Of course, many will choose to believe that they are against my thesis because the arguments against it are so compelling. Precisely which are those arguments? Or is their more basic motivation perhaps the fear that a science of human action would rob us of our dignity or threaten our human autonomy? Are they afraid that a science of human behavior would somehow dehumanize us, and are they therefore motivated to find an argument against it? It is the last response that I have characterized as resistance to knowledge. It occurs when we have a vested interest in the outcome of empirical matters, as we do in the debate over a science of human behavior, and allow ourselves to react to it emotionally rather than rationally; it occurs when we think that we have more to fear from the pursuit of truth than from our ignorance.

Resistance to knowledge is not unique to the social sciences. Indeed, it is prevalent in many areas of inquiry, especially those in which scientific scrutiny would seem to threaten our cherished political or religious ideologies. All of the greatest scientific advances have represented profound revolutions in our beliefs and restructuring of our metaphysical assumptions. The crime of Galileo was not that he discovered craters on the moon or satellites orbiting the planet Jupiter, but that by doing so, he threatened the Aristotelian dogma that the heavens are permanent and perfect, as fixed by God. Similarly, the Darwinian revolution in biology can only be understood as an affront to the view that humans are at the top of the great chain of being. Of course, with perfect hindsight it is easy to suppose that each one of us would have been able to see the sense in each of these intellectual breakthroughs—but partially this is because our own lives and educations have been weaned on them. Yet what backward and unwarranted superstitions and beliefs do we today harbor as a result of our cultural and social upbringing? Only a fool supposes that he or she was luckily born into the generation in which all truth had finally been revealed—and yet how many of us, in our daily lives, act as if this were true?

Resistance to knowledge occurs today about contemporary empirical matters in exactly the same way as it did in each of the historical debates cited above. And the test of whether you are beholden to resistance to knowledge is not whether you now think that if put back in time you would have agreed with Darwin or Copernicus. When all of the facts are in, it is easy to be on the side of reason. Rather, ask yourself honestly how you would judge each of the following matters of current scientific debate and whether you are willing to say to each: "It is an empirical question":

- Whether human behavior can be explained primarily as a function of biological factors
- Whether there is a link between race and IQ
- Whether artificial intelligence will ever produce a machine that is capable of creative thought
- Whether homosexuality is genetically predisposed
- Whether resistance during a rape attempt is more likely to lead to injury or prevention of the attack
- Whether a victim's manner of dress incites a rapist
- Whether there are gender differences in reasoning
- Whether Hitler's foot-soldiers were exceptional in their willingness to follow orders, or whether we all are capable of committing atrocities under the right circumstances
- Whether the death penalty deters crime
- Whether day care has a detrimental affect on the cognitive development of children
- Whether spanking a child causes him or her to become more aggressive later in life
- Whether our behavior is predictable
- Whether we have free will
- Whether it is possible to have a science of human action

My intention in presenting this list is not to suggest that one side or the other in these debates is correct. Nor is it even to suggest that uncongenial ideas are more likely to be true merely because they have been neglected. It is simply to challenge readers into asking whether they are willing to adopt a scientific attitude toward the study of questions that are empirical. Are you willing to say that the only relevant factor in deciding the above debates is scientific evidence? Are you willing to put your prejudices to the test, and to formulate your beliefs only on the basis of the evidence for them?

The answer given by the majority of social scientists, unfortunately, is that they are not. Of course, no good social scientist readily admits this. But what they do is reveal their prejudices by radically overestimating the barriers to the empirical study of highly charged ideological matters. Why? For the same reason that each one of us individually might react to any of the above listed issues by saying, "I hope the answer isn't yes" or "I don't want to know the answer." *It is because we fear the truth when it does not already match our prejudices.* When we are afraid of what we may find out, we often react by refusing to believe what evidence has come forward or by resisting further investigation. Yet this reaction is antithetical to the scientific attitude. And if it is not appropriate in our investigation of nature, why suppose that it is any more appropriate in the investigation of our social relations?

At base, the problem in social science is that one side or another of each of the above debates runs afoul of the political, religious, or egocentric prejudices that each of us has about who we are as human beings. Of course, we cannot help but have vested interests in matters that affect us so directly. And it is fruitless to ask us to suspend our judgment until all of the facts are in. But what we can help is whether we will allow our interests to chill our willingness to investigate the empirical facts about matters where we are perfectly capable of doing so, and so whether we will resist knowledge that threatens our worldview. Are we courageous enough to face ourselves directly, without the veil of superstition, myths, and lies that we construct for ourselves to make life more palatable?

We must get over the prejudice that truth will be easy to obtain and beautiful to behold. Neither Copernicus's nor Darwin's revolutions should lead us to think that restructuring our worldviews will be easy. So why should we suppose that the truth about human nature will be pretty, and will easily accommodate our political or religious ideologies or our everyday intuitions? Are we willing to believe that humans are cruel by nature? That we are perfectly determined creatures devoid of free will? If we are not, if we are willing to put intuitive limits on what our social investigations may discover, how can we pretend that the fruits that they bear will be those of a true social science?

Of course, my argument here should not be taken as an endorsement of any particular outcome in matters of social scientific debate. If they are truly empirical matters, then it would be equally foolhardy to rely on our counterintuitions as it would be to rely on our intuitions. Whether any of the above matters are true is a matter of empirical fact. But what does seem to be sanctioned is a sort of mental toughness toward undoing our prejudices about human nature, in service of the scientific attitude toward empirical evidence. That is, we should be prepared to be hard, or even harder, on our cherished beliefs as we are on those that we would find it inconvenient to believe. Why? Simply because over the history of scientific inquiry, we can count on the fact that our subjective prejudices have been at work and have biased our investigations. The things that are convenient to believe—that we are altruistic beings by nature, that Hitler's soldiers were exceptional in their compliance to their leader, and that there are no differences in spatial and verbal reasoning between the sexes—do not need advocates who are willing to be burned at the stake for what they believe. They are knocking on an open door. Rather, it is the horrible ideas—that we are genetically or instinctually determined beings, that there is nothing exceptional about "life" or "mind" that separates us from other matter in the universe by some supernatural act, or that we alone are in control of human destiny—these are the ideas that deserve our closest scrutiny. For a true science of human behavior, we must be willing to investigate not only those theories that are attractive but also those that are repugnant.

Few would like to believe that the truth about human nature is not congenial to us. But perhaps all of the easy truths have already been discovered. Are we willing to face the horrible conclusions that a truly rigorous social science might reveal to us? Ask yourself what you do not want to believe and be willing to search there for the truth about human nature. Be willing to follow the truth into the dark alleys and avenues, pursuing most doggedly what we do not want to believe about ourselves—that is where a science of human behavior must be willing to begin.

Perhaps the truth about us will not be horrible, but will be uplifting and inspirational. We will not know it unless we are willing to apply the same standard of evidence in examining our fears as we have thus far in confirming our hopes about human nature. We must be willing to put our trust in empirical procedures when dealing with empirical questions and let our attitudes be shaped by science and not by ideology. Few people are averse to giving their cherished beliefs the benefit of the doubt. But how many of us are willing to discount our prejudices in favor of empirical investigation? Yet this is what is required by science.

What is most troubling to realize, however, is that it is this very attempt to be rigorous that is at the root of most people's objections to a science of human behavior, for many fear that the outcome of a science of human behavior could not help but be awful. This prejudice, when I first encountered it, surprised me. For one reason or another, I hadn't expected the reaction against the idea of a science of human behavior to be so uniform and so predictable. But my experience in defending a science of human action has been such that I have come to expect that nearly everyone reacts emotionally to it at first. There are few who do not have an opinion. By and large, for some reason I have never quite been able to fathom, most human beings do not want to be the subject of scientific inquiry.

Years ago, when I was first formulating my ideas on this subject, the response I got from virtually every liberal academic was axiomatic. They compared the desire to have a science of human behavior with *Brave New World* or even to Nazi Germany. (I have since come to expect a similar reaction from libertarians, who have their own axe to grind about determinism). The philosophical and methodological arguments came later—the emotional reaction came first. But why? What precisely is at issue that provokes such a strong unilateral human response? Why is it so upsetting to think that we could have a science of human behavior?[1]

The threat, it seems, is one to human freedom and autonomy—the notion that we are "special" in the universe. It is taken to be an insult to human dignity to suppose that one can study our behavior with the same methodology that one uses to study all of the other matter in nature. It degrades us and robs us of our uniqueness. It is somehow to make us less than human. Jacob Bronowski, in his important work *The Identity of Man*, captures well what is at issue. He writes,

> What is it that troubles us in the assertion that living things are made from the same atoms as dead, and ruled by the same physical laws? We may pretend that our difficulties are intellectual, and that we are merely puzzled how this could come about. But our uneasiness lies deeper. It lies in a feeling that if the dance of atoms in our bodies is not different

in kind from the pattern in the star and the stone, then we have suffered some loss of personality; a denial of mind in our sense of human self.[2]

But what is this feeling but mere prejudice? Does it have any empirical support? Does it have any scientific basis whatsoever? And were not similar prejudices once at work in governing our beliefs about the position of the earth relative to the sun or the place of human beings atop the animal kingdom? Why should we give any cognitive weight to the intuition that human beings are "special," and so not subject to the methodology of science? Yes, this belief seems to be part of the human package, and it is widely shared across the members of our species. Indeed, the extent to which such a belief is self-serving probably explains why it is so widespread. But does this make it any less likely to be true that our behavior is in fact governed by a set of laws that are discoverable by scientific procedures? Certainly not.

The most probable reason that most people are not willing to accept a science of human behavior—whether they admit it or not—is that it runs afoul of the major tenets behind our dominant religious and political ideologies. But these, of course, spring most basically from the egocentric belief that humans are special, and thus that our behavior is not amenable to scientific treatment. We become hostile at the prospect of someone telling us that they know why we act as we do, with the implication that we do not! And as I have already suggested, this view is incompatible with the religious and political myths we invent to legitimize our superstitions and salve our ignorance about ourselves.

But as I shall now argue, it is this very reliance on ideology that is the largest source of resistance to knowledge. For what is ideology but the conviction that we already know, before empirical investigation, what is the case? Ideology stood in the way of natural science in its day, as it now stands in the way of empirical social science. Indeed, religious and political ideology are two blind alleys into investigating the true causal factors behind human behavior and, I maintain, stand as the biggest impediments to a science of human behavior. It is these twin views that are largely responsible for legitimizing the prejudice that we can manage human affairs without precise knowledge of the causal forces behind our behavior, either (in the case of religious ideology) because we believe that the true cause is God or (in the case of political ideology) because we believe that any causes can be overriden by normative considerations. Let me discuss each in turn.

Religious Ideology

Insofar as our religious beliefs are based on faith, religion is antithetical to the scientific attitude, which is based on doubt and empirical evidence. Where they have clashed over matters of empirical fact, religion has been offered as a substitute for reason and has retreated only after suffering the most crushing defeats. These defeats have come at the hands of science.

Over the centuries, there have been many attempts to reconcile religious belief with scientific practice, and some have even been moderately successful. But where they cause us to rely on faith rather than sensory evidence in the examination of empirical matters, they come at the cost of compromising scientific integrity. Indeed, in Christianity we see most plainly the source of the fundamental objection to a science of human behavior. It is that science is "humanistic."

Humanism, which traces its roots back to the Enlightenment, is a system of belief that upholds the idea that humans are directly responsible for the condition of their existence on earth and that we are capable of improving our lot through rational thought, without appeal to supernaturalism. It is clear what Christians find objectionable about this doctrine. For in their view, we are taking things into our own hands that properly belong in God's. We are making our fate our own responsibility. We are focusing too much attention on achieving improvement in this life, when we should instead be preparing ourselves for the next one.

It may seem tempting at this point to conclude that science and religion are mutually incompatible. After all, how can anyone who purports to believe things based only on rational consideration after gathering empirical evidence believe in a concept so patently implausible as God? I do not myself believe, however, that where scientific practice is concerned the opposition need be so severe. True, if the claim behind Christianity is that God really exists, then that too is an empirical question. But further questions are then raised concerning what one is willing to accept as evidence, and what one is willing to call rational on the basis of that evidence. And such a radical opposition between science and religion must confront the fact that natural science has prospered even at the hands of investigators who held strong religious beliefs, such as Copernicus, Newton, and Einstein. Rather, the appropriate response seems one of true agnosticism, whereby one resolves not to let religious answers be given to scientific questions and also not to let science address theological matters. Thus, in our social scientific investigations, as in our inquiry into the natural world, we must resolve that religion is no substitute for science.

The attitude that social scientists should take to religious belief therefore may be much like that which I proposed to free will in chapter 2. It is an empirical question whether God exists, and either way we need to do the best that we can to improve our condition on earth while we are here. If the laws of human nature are as beautiful and seem so divine as those of nature, let us discover it for ourselves. We cannot wait for God to provide the answer. For if God exists, he seems to care little about this world and worries mainly about the next one, else why does he allow torture, rape, and starvation? Or perhaps if God exists, he is not all that we have imagined. As far as social science is concerned, it does not matter. Whatever God's plan, the social sciences should be concerned with improving the lot of human beings while they are living on this planet and should not be distracted with theological matters.

Social science is about *us*, not about God. And it seems clear from the history of human life over the millennia that we are directly responsible for the condition of our existence here on earth. (Indeed, such a position is easily reconcilable with the religious response to the "problem of evil," which tells us that evil is the fault of human misuse of free will, and not of God. If evil is a result of our misuse of free will, what better employment for social science than to tell us how to use our free will toward better ends?) Moreover, just as in the study of nature, the scientific study of human behavior is not the study of how the laws that govern our action came into being, but only how they work. Just as the physicist chooses not to inquire into why there is gravity, but studies only its behavior, let us also admit that the study of how we function is the proper province of social science, leaving it as a theological or philosophical question how we came to be the way that we are.

But one thing is certain: metaphysical speculation should not take the place of the empirical investigation of matters about which there are empirical data. In such cases, the clash between scientific and religious belief is a result of religion overstepping its domain; to the extent that it continues to follow the path of deciding empirical questions on nonempirical grounds, religion will continue to suffer more defeats like those it has been handed by the scientific theories of Copernicus, Galileo, and Darwin as social science advances. Indeed, it is unclear why empirical matters should be the concern of theologians, any more so than scientists should try to decide spiritual questions. May not those with theistic beliefs resolve to allow social science to explore human nature separate from religion, and even to encourage it to formulate its own conclusions?

Thus, just as the investigation of nature or the practice of technology is not seen to be impious, let us so take it as our task to understand and to do something about our condition here on earth, without being beholden to religious ideology. And let us realize that while the debate about the relationship between science and religion may long continue, the empirical facts about human behavior, as well as those about nature, should be considered the exclusive domain of science.

Indeed, as we shall see in the next chapter, the principle that the Bible is an authority only in spiritual matters, and not on empirical questions about nature, may be called the "Galilean principle," for it was used by Galileo 350 years ago as the only means to protect religion from erosion by the advancement of science, and scientific independence from religious ideology. That the Catholic church notoriously rejected this principle, imprisoned Galileo, and suffered defeat after defeat, until it finally accepted it in the nineteenth century, is the topic of my next chapter. What I am recommending here is that those religious ideologues who have strayed from the Galilean principle recognize its force in preserving not only science but also religion, and consider extending it to include all scientific investigations, including those beyond nature. If the Galilean principle is now accepted by the church as governing its relationships with natural science, why not also with social science? The benefits of doing so are identical to those originally urged by Galileo. In the long run, one may certainly foresee that without it, religion will once again be forced into retreat in the face of the empirical advance of social science.

Political Ideology

I have met many atheists and agnostics who embrace a liberal political philosophy and would agree with every word I have just said against the corrosive effect of religious ideology on scientific investigation and would regard themselves as exempt from the corrupting influence of such ideological dogma. In my experience, this is rarely the case. There are true believers aplenty both within religion and without. Just because we see the dangers of one ideology is no guarantee that our reason is not being corrupted by another. What I am arguing here is that reliance on an ideology of any kind, whether religious or political, whether liberal or conservative, is dangerous because it depends on the assumption that we can in some matters substitute conviction for reason—that we may sometimes trump or ignore empirical evidence in the service of normative or spiritual considerations.

The problem with ideology is that it runs afoul of scientific skepticism and proceeds instead from the belief that we may make sense of our plight based on an understanding of the facts about our existence that, in the fullness of time, has every chance of turning out to be wrong. In short, I believe that where it is not constantly subject to revision by continual comparison with empirical evidence, political ideology can be just as prejudicial and destructive of the scientific attitude as religious ideology.

For proof of this, one need look no further than the current wave of political correctness that has swept our colleges and universities. Like religious dogma, the mind-set of many of those whom we might expect today to be doing empirical social science is fixed not by the standards of science, but instead by the reigning pieties of political correctness. There have been written whole volumes on the corrosive power of political correctness for rational thinking. The temptation to decide empirical questions on the basis of their political consequences, the conviction that inconvenient facts can be subjugated to normative considerations, the reluctance to investigate issues that run contrary to current political opinion, or even the conjuring of empirical fictions based on nothing more than wishful thinking—these are some of the manifestations of political correctness.

To anyone who has not recently spent time on an American university campus, it is probably unimaginable that rational discourse on some of the most important questions of our age has now been taken hostage by self-appointed "thought police," who are ready to find racism, sexism, and classism at every turn, even in the gathering of empirical data. Yet in an unguarded moment, any social scientist will tell you that it is true. And they will also tell you that the consequences of ignoring this ideology are severe and can result in public ridicule, denial of tenure, or ostracism. That such a pervasive and dogmatic ideology could not help but to chill free inquiry into the empirical facts of human behavior can be denied only by those who are ignorant or who themselves are in denial over the dangers of ideology for scientific inquiry.

What exactly *is* political correctness? Even to attempt a definition is to take sides in the debate. To its advocates, political correctness is a perhaps unfortunate label given to those who are concerned with defending the rights and sensitivities of the downtrodden from undue prejudice exerted by those who are in power. To its critics, political correctness is a rigidly ultraliberal political philosophy whereby Western culture in general (and white males in particular) are made to blame for all of the inequities in life suffered by ethnic minorities, women, homosexuals, and the disabled. It is condemned by its foes as a doctrinaire political ideology that is severely intolerant of even the slightest deviation from orthodox liberal positions.

In its incarnation on college campuses, advocates of political correctness have been particularly concerned to root out "oppressive speech" and have sought to displace a vocabulary that is seen as implicitly racist and sexist with one that is more neutral. Thus, "freshman" becomes "first-year student," "black" becomes "African American," "disabled" becomes "physically challenged," and, in some places, "women" becomes "womyn" (for fear that the traditional spelling denigrates women by making their very name a derivative of the word for "men"). On many campuses, conformity with such rules has been dictated by the imposition of "speech codes," whereby those who violate the use of these rules are presumably racist or sexist and are correspondingly punished with sometimes severe university sanctions. Ever vigilant for new forms of oppression, one also hears these days of those who are condemned for "lookism" (preferring those who are attractive to those who are not), "speciesism" (preferring humans over other species), "ablism" (preferring those who are "temporarily able" to those who are disabled), "heterosexism" (preferring heterosexuals to homosexuals), and "ageism" (preferring those who are young to those who are old). Determined to root out any possible grounds for prejudice on the basis of inequality, the net effect of political correctness has been to reflexively attack any mode of thought that appeals to inherent differences between people as grounds for preference between them.

Of course, it would be wrong to make it seem as if political correctness, any more so than religious belief, is one coherent set of doctrines, from whose fundamental principles one can derive a fully consistent set of compatible beliefs. Yet there is an orthodoxy of content behind political correctness, if not of principle, that gives us ready political answers to empirical questions. Such convictions are meant to forestall the need for empirical investigation into taboo subjects, where we may turn up findings that undermine our egalitarian beliefs. And this is what I have claimed is dangerous about letting political correctness, or any other ideology, infect our empirical investigations. Where ideology is accepted as a suitable substitute for scientific inquiry, we are encouraged not to investigate matters that may be of enormous importance for improving our social problems. Sadly, many of today's social scientists willingly accept the idea that there are certain topics about which the gathering of empirical data is taboo.

Perhaps the best and most recent example of this hostility of political correctness to the investigation of unsavory empirical topics can be seen in the academic reception of *The Bell Curve*. In this work, Richard J. Herrnstein and Charles Murray argue, among other things, that there are demonstrable empirical differences in IQ across different ethnic groups. That such a conclusion runs afoul of the dogma of political correctness is obvious. Indeed, the reaction to *The Bell Curve* has been largely political. It has been called "the most incendiary piece of social science to appear in the last decade,"[3] and its conclusions have been denounced as "frightening." Indeed, several commentators have claimed that what is basically wrong with *The Bell Curve* is that it presents empirical justification for the unequal treatment of those from different ethnic groups. Several anthologies of criticism have appeared, in which many of the essays attempt to show that *The Bell Curve* is nothing more than a thinly disguised attempt to further a racist political agenda. But what of the status of the empirical claims that Herrnstein and Murray made? Are they nonetheless true?

Rather than excitedly running to their labs for further empirical inquiry, the predominant reaction of most academics has been to denounce the politics of the work. Missing, however, is the realization that as a piece of social science, *The Bell Curve* must be evaluated not on the basis of its politics but on whether the facts it presents are accurate. Indeed, if the facts in it are wrong, nothing would make shorter work of them than empirical refutation. The response of most social scientists to this challenge, however, has been given over to cheerleading for political correctness.

It is significant to note, however, that there have been responsible empirical evaluations of the arguments presented in *The Bell Curve*, that have revealed serious methodological shortcomings in this work.[4] And yet the majority of social scientists have been satisfied to dismiss the work on political grounds and have taken little notice of these empirical findings. Indeed, rather than calling for more empirical investigation to settle the dispute now left open by the alleged methodological flaws in Herrnstein and Murray's work, as happened in the debate over cold fusion, one senses a collective sigh of relief among contemporary social scientists that now that the odious political assumptions behind *The Bell Curve* have been uncovered, they fortunately do not have to deal with the topic any longer. This is a sad commentary on the current state of social inquiry and gives little hope for its future as a scientific enterprise as long as it lives under the ruling hand of political correctness.

Of course, it would be wrong to suggest that the only political biases in social scientific research are from the left. Biases from the right are no better and are no less dangerous to the pursuit of a genuine social science.[5] One might consider, for example, the conservative political belief—despite overwhelming empirical evidence to the contrary—that the death penalty deters crime. Of course, it is true that the death penalty succeeds at what criminologists have called specific deterrence, in which the individual perpetrating the punished act is deterred from ever doing so again once he is put to death. Rather, what has been at issue is the success of the death penalty in fostering general deterrence, whereby other individuals are deterred from performing a similar act when they learn the fate suffered by another. Despite ample empirical evidence that the death penalty does not succeed in this, many conservatives have been reluctant to give up their belief that the death penalty promotes general deterrence.[6] But as with political correctness, need we search overmuch for the source of such strong convictions, in the absence of empirical verification? No. One might reasonably conclude that the source is ideological; it is based on the extraempirical political belief that the death penalty *should* deter criminal behavior, and so it must in fact do so.[7]

What are we to conclude about the overall effect that political ideology has had on empirical social scientific research? It is my thesis that its influence has been profoundly negative and that it has chilled the pursuit of the scientific attitude in social science. At times entire areas of research have been ruled out of bounds simply because their investigation may uncover facts that cause us to question favored ideological assumptions (or violate the basic human prejudice that humans are somehow special). In some instances, this has led to the suggestion that social scientists cannot settle an empirical dispute, simply because it raises political issues that are "controversial." In other instances, well-confirmed facts about human behavior have been rejected altogether simply because they contradict popular political ideals. In this way, the corrosive effect of political ideology reveals its similarity to other age-old attempts to resist knowledge, seen throughout the history of scientific inquiry. Indeed, it is my contention that political ideology is today doing to social science what religious dogma did to natural science in the first Dark Ages.

The fruit of this analogy will be more fully appreciated after reading my next chapter, in which I present the details of how it was that religious ideology came to clash directly with the scientific attitude in the time of Galileo. Of particular importance now, however, is to realize that what I am arguing against here is all ideology, insofar as it hampers the spirit of scientific inquiry. To those who reject the corrosive influence of religious ideology yet embrace the fundamental tenets of political correctness, I hope to have shown the dangers of even the best-intentioned meddling with empirical inquiry. And to all who would seek to subject the practice of social scientific inquiry to the scrutiny of our political agendas, I hope to have demonstrated the danger that this poses to true scientific inquiry.

One might now reasonably ask, however, in light of all of these political pressures from the left and from the right, whether it is reasonable to think that good social science can be done at all. The answer, I submit, is a resounding yes, for despite such ideological pressures, good empirical research is currently going on in social science.[8] Just as Galileo and Darwin fought the reigning religious ideologies of their day and embraced the scientific attitude in gathering empirical data, so today there are examples of social scientists who have fought enormous political pressure from both the left and the right, and have insisted on sticking by the scientific attitude in their work. Such an example is Gary Kleck's work on guns and violence, which emphasizes the gathering of data and the interpretation of trends based on empirical evidence, leaving policy recommendations to be made only once we have the facts before us.

An Example of Empirical Social Science

Discussion of the relationship between guns and violence usually provokes a strong emotional reaction in most people, and the majority of social scientists have proved to be no exception. Indeed, until relatively recently, the majority of work on this topic was marred by those who had a particular political agenda to advance, and the credibility of the studies cited was questionable, as disconfirming studies were often ignored and extensive empirical work had yet to be undertaken. In the absence of objective empirical analysis, several myths grew up surrounding the relationship between guns and violence, most of which served the political agenda of gun control advocates. Three of the most prevalent beliefs were that:

1. Since many violent crimes in the United States involve use of a gun, banning guns would cut the rate of violent crime.

2. Those who use guns to defend themselves against a crime are more likely to have the gun taken away and used against them than to prevent the crime itself.
3. A gun in the home is many times more likely to be used accidentally against a family member than an intruder.

In 1991, Gary Kleck, a criminologist at Florida State University, published the definitive study of the relationship between guns and violence in America, *Point Blank: Guns and Violence in America*. This massive volume reviews all prior statistical and survey data in the field and supplements it with original studies, where appropriate. In contrast to much of the prior literature available in the field, Kleck's work is exclusively empirical and offers no political, emotional, or other sorts of appeals on one side or the other of the gun control debate. Indeed, as Kleck points out at the beginning of his book, although he is a lifelong Democrat and a member of the American Civil Liberties Union, he has received no funding from either side in the gun control debate and strives solely to gather empirical data. Save a few policy evaluations at the end of his work (once the evidence is clear), *Point Blank* is entirely empirical in its focus and treats the subject of guns and violence as an important topic for objective social scientific research.

The details of Kleck's findings are remarkable. With reference to the three beliefs cited above, Kleck has found that:

1. Americans use guns to defend themselves against crime as often as a million times a year, over 600,000 for handguns alone. Given estimates of criminal gun use of approximately 540,000 for handguns and 660,000 for all guns, in any given year, the evidence indicates that "defensive use" of guns is roughly equal to that of criminal use.[9] Thus, any crime-inhibiting effect of gun

control legislation should be weighed against the crime-enhancing effect of eliminating defensive gun use.

2. Contrary to popular opinion, victims who respond to criminal confrontation with a gun are much less likely to be hurt than victims who respond in any other way, including not resisting at all. Indeed, based on National Crime Survey statistics, it was found that at most, 1 percent of defensive gun uses resulted in a criminal taking a gun away from a victim.[10]

3. Fewer than 2 percent of fatal gun accidents involve a person accidentally shooting someone mistaken for an intruder.[11] The total number of other fatal gun accidents (FGAs), moreover, is very small (approximately 1,500 annually) and largely confined to a small share of households with members willing to engage in reckless behavior with guns. Given the large number of defensive gun uses (DGUs) and the relatively small number of FGAs, it is not in fact true that a gun in the home is more likely to be used against a family member than an intruder. (It is worth noting here that those who have argued otherwise often rely on a faulty methodological ploy where one measures the number of FGAs against the number of intruders killed. But surely that is to radically underestimate the total number of DGUs, the majority of which are not fatal).[12]

Although numerous other results are reported in Kleck's work, these three findings have been the focus of vehement attacks on his work by gun control advocates. This is not to imply that Kleck's work has been successfully challenged on empirical grounds. To the contrary. It has been hailed as the definitive study of guns and violence, and in 1993 *Point Blank* received the highest award of the American Society of Criminology. Rather, it has been attacked largely because the results it turns up are so uncongenial to the political beliefs of gun control advocates.

Perhaps the most vehement reaction against Kleck's book has been that of those working in the field of public health. Indeed, in his review of the reaction against Kleck's work, Don Kates and his coauthors find that Kleck's findings are often suppressed, and that where Kleck is cited, it is often erroneously stated that his views have been "discredited," or, worse, he is dismissed as a "Neanderthal gun supporter."[13] Kates cites numerous other instances in which authors cite Kleck's critics, or other studies that Kleck has successfully challenged, but not Kleck himself. Such ploys are indicative of resistance to knowledge at its worst.

It is interesting to note, moreover, that Kleck's work has also been attacked by those who are *against* gun control, given his conclusion at the end of the book—based on his empirical findings—that the most effective means for the prevention of criminal use of guns would be a form of limited gun control (involving extensive background checks, registration of gun owners, and strict regulation of the transfer of guns) that would not conflict with defensive gun use by law-abiding citizens. Indeed, Kleck's support of a form of gun control much more sweeping than (and predating) the Brady Bill has made him no darling of the gun lobby. Thus, we see in Kleck's work an interesting duality; his insistence on gathering empirical data on such a hot-button political issue, and then basing policy recommendations on these findings, has resulted in his work being attacked by political ideologues on both the left and the right.

There has also been some responsible criticism of Kleck's work by those who question a few of his empirical findings. Specifically, Kleck's primary critic, Duke University economist Philip J. Cook, disputes Kleck's figure for annual defensive gun uses, putting the figure somewhere over 100,000. But for the purpose of my argument here, who is right in this debate does not matter.[14] What I am arguing here is not that Kleck's work is right in all of its findings or even that there could not be some previously undiscovered flaw in some of his methodological procedures. These, after all, are empirical matters. Rather, what I am arguing is that Kleck is right that the debate about guns and violence is an empirical, and not a political or an emotional, issue and that it deserves to be decided on the basis of the evidence. Thus, whether he is right about all of the specific findings in his book or not, I find much to admire in Kleck's work and think that there is a lot to be learned from it as a lesson for those who wish to construct a science of human behavior. In *Point Blank* Kleck embraces the scientific attitude and bucks the forces of resistance to knowledge. He effectively demonstrates that despite strong ideological resistance, good social scientific work can be done.

I hope it is also clear by now that what I am arguing against is the efforts of those who reject Kleck's work primarily because of their emotional stance on guns and who desire to subordinate the empirical facts about the gun debate to their political agendas. Indeed, in the debate about guns and violence, we can see at work some of the most virulent aspects of political correctness in its fight against a science of human behavior. As an all-encompassing ideology, political correctness embraces the assumption that we have nothing much to learn from the empirical study of human action if it does not already match our intuitions. Thus, if the empirical study of guns and violence turns up findings that undermine the political belief that guns should be banned, the empirical findings are called into question or rejected outright. Is this not precisely the reaction of many of Kleck's critics? And is this not the very antithesis of the scientific attitude, so needed to launch a true science of human behavior?

In Kleck's work, we see a clear example that a science of human behavior is possible, despite the ideological assumptions and outright prejudices that humans bring to the study of their own behavior. And in the reaction to it—in particular, the antiempirical and knowledge suppressing follies of political correctness—we witness the danger of allowing even the most attractive ideologies to have a place in our scientific disputes.

I have said enough by now to offend almost every one of my readers. Fear and rejection are common responses when our prejudices are questioned, especially about matters that concern us so deeply. But what we need now is courage in the face of the task before us. Ideology—either religious or political—has not worked in improving the quality of human life, and indeed at times it has legitimized some of the most horrible crimes that human beings have committed against one another. It does us no service to believe things, and base our social policies on them, if they are not true. This is the road to ruin, and it will only lead to a worsening of the human condition. The only thing that can rescue us is a true science of human behavior.

Ideology makes it convenient to believe what we want to believe in the absence of evidential confirmation. Unlike rival theories, rival ideologies are not decidable on the basis of empirical investigation. And yet the primary feature of a science of human behavior that religious and political ideologies have found offensive is the idea that we should put our intuitions about ourselves to the test in case we are wrong about them.

Would the banishment of plea bargaining lower the crime rate? Does raising the minimum wage contribute to unemployment? Are legal immigrants more likely to be on welfare than native-born Americans? Why do we bother to speculate on these matters from the platform of our ideologies? The answers to all of these questions are empirical and therefore are provided not by ideology but by scientific investigation.

But we now face a difficult issue, for the topics we have been discussing *are* emotionally charged, and we *do* care how such questions are answered because they are directly about us. We cannot help but have our own opinions, informed or not, about these matters. And even if this does not amount to proof that we cannot have a science of human behavior, as we learned in chapter 2, isn't this nonetheless a formidable barrier to social scientific inquiry? Is there any hope of objectivity? Can the scientific attitude thrive in an environment of such ideological turmoil? Isn't social science a special case because we have so many, and such deeply vested, interests?

Such questions deserve our deepest consideration. And yet the challenge of answering them is one that has been faced, and met, by science throughout its history. For although it is easy to forget, these very same barriers of resistance to knowledge, fueled by ideology and superstition, have shadowed scientific advancement from its inception. Yet science, when practiced correctly, has triumphed.

The answer to the question of whether science has any hope of providing empirical answers to questions that are fraught with religious and political considerations is a resounding "yes"! For in the ideologically charged minefield of the history of natural science, we shall see that natural science once faced a degree of resistance to knowledge that is even more chilling than that now faced by social science.

4 A Lesson from the History of Science

The Bible tells us how to go to Heaven, not how the heavens go.
—Galileo, 1615

I, Galileo, being in my seventieth year, being a prisoner and on my knees, and before your Eminences, having before my eyes the Holy Gospel, which I touch with my hands, abjure, curse, and detest the error and the heresy of the movement of the earth.
—Galileo, 1633

My goal in this chapter is to examine what the history of natural science has to teach us about the future of social science. To what extent are the situations similar? Is it fair to say that social science is now in its Dark Age? What were the Dark Ages of natural science like? It is tempting today to indulge in the fantasy that natural science had it easy—that not only did it not suffer from the sort of methodological barriers outlined in chapter 2, but that the truths of natural science were so obvious that their discovery was inevitable even in the face of ideological resistance. Indeed, from today's viewpoint, the ideological resistance to natural science may seem ridiculous. How could that situation be our current one in social science?

This view, however, is mistaken, and is known to be mistaken by anyone who has made a serious study of the history of natural science. There has been (and in some corners continues to be) incredibly strong ideological resistance to the advances of natural science. Indeed, in its own Dark Ages (and in some cases far beyond), natural science faced many of the same sorts of barriers that now confront social science. Clearly, I am not here talking about similarities in the empirical details, but rather the clash between the scientific attitude and the forces of resistance to knowledge. Today the battle lines are different, but the war is the same.

Natural science never had it easy. There were no disinterested observers waiting patiently in the wings for the verdict of science, for this is not how the human mind works. We do not suspend our judgment in the hope of someday codifying our beliefs only when we can do so on a rational basis. When we do not yet have good evidence for our beliefs (and sometimes even when we do), the void is filled by superstition, myth, and ideology. Indeed, though it is now forgotten by most observers, science once lived under the ruling hand of theology and Aristotelian philosophy and could venture no explanations that did not accord with the reigning ideology put forth by the Roman Catholic church. Natural science (which was then called "natural philosophy") broke free from this only relatively recently (in the seventeenth century), and only then because of the courageous convictions of its practitioners. There was bitter resistance to the founding of modern natural science, when it began to put forth opinions that clashed with the ideology of the church, and there was much suffering because of it. Giordano Bruno was burned at the stake in 1600 for his uncompromising belief in other worlds. Galileo was put under house arrest in 1633 and forced to abjure his belief in the motion of the earth. The forces of resistance to knowledge were alive and well in the history of natural science. Indeed, such examples should make current social scientists blush for their own lack of courage in the face of public ridicule and hostility to unpopular ideas. Although resistance to knowledge may now work in social science in much the same way as it always has, the consequences today are much less dire than those suffered, for instance, by Galileo.

Galileo

Contemporary social science has much to learn from Galileo. In him we see the embodiment of the scientific revolution in full flower, at the defining moment when science broke free to become an independent discipline. What is so important about the example of Galileo is not just the idea (Copernicanism) that he was fighting for, but his willingness to fight for it with such vigor against the forces of resistance to knowledge brought to bear by the Catholic church. In the story of Galileo, we see in sharpest focus that natural science too faced incredible resistance to knowledge. Yet perhaps even more important, we also see in Galileo the foundation of the modern conception of science. No longer satisfied merely to rely on the authority of Aristotle or the church—where one is obliged to check the facts about nature only against our reason or our faith—Galileo ushered in a new method of scientific practice that was experimental. The power of these two ideas—that science may stand up to the forces of resistance to knowledge, and that it may decide matters of fact solely on an empirical and experimental basis—is the reason that many consider Galileo to be the father of modern science.

Yet, having said this, I nonetheless feel obliged to defend my choice of Galileo as an example for modern social science. For surely, as some will point out, although Galileo fought valiantly for a scientific idea—that the earth moves—it was Copernicus's idea that he was fighting for. Likewise, although Galileo did much to foster the experimental attitude in natural science (in contrast to the sterile musing of scholastic philosophers), it was Roger Bacon (and his teacher, Robert Grosseteste) who first put forth the idea that science should be based on sensory evidence, and is therefore empirical, and Francis Bacon who later codified this into a well-known methodology. So why Galileo?

I have chosen to focus on Galileo because in him we see personified those qualities without which the emergence of modern science might have been indefinitely delayed or might never have occurred at all. It was not just Copernican theory that Galileo was fighting for, but the existence of science as a separate discipline. Before the time of Galileo, science was subordinate to philosophy, which was in turn subordinate to theology. In Galileo's stand against the Catholic church, we see a scientist with the courage of his convictions, who was willing to stand up to the forces of resistance to knowledge. Thus, it was not Galileo's belief in Copernican astronomy that is so important as a lesson for the social sciences but the fact that he was willing to fight for it. Moreover, although it is true that the empirical method of science was not invented by Galileo, it was he who tempered it with the importance of mathematical insight, popularized it, and so lived it in his scientific achievements as to make it fully his own. Thus, although one may quibble over Galileo's scientific reputation as a theoretician or even as a methodologist, his embodiment of the scientific attitude, and his willingness to defend it, forever changed the relationship between science and religion. It was Galileo who set modern science on the right footing, and the result has been the enormous success that we now take for granted. Yet although the story of Galileo is one of the most important moments in the history of natural science, its details are largely unknown to the social sciences.

The facts of Galileo's trial before the Roman Inquisition are these. Far from being a life-long Copernican zealot, as is commonly supposed by those unfamiliar with Galileo's work, Galileo instead came to be convinced of Copernicus's theory (which says that the Earth along with the other planets orbit the sun, rather than that the sun and the planets orbit the earth) slowly over time, by a matter of degrees. Probably crucial in his conversion to this theory were his own telescopic observations, starting in 1609, that resulted in several startling discoveries, which fit neatly with Copernicanism but ran contrary to the prevailing geocentric theory. Single-handedly Galileo discovered four moons of Jupiter, craters and mountains on the moon, sunspots, and the phases of Venus. Each of these discoveries in its own way was devastating for the Ptolemaic/Aristotelian system then favored by the church, which supposed not only that the earth was at the center of the universe but also that the heavens were perfect. Without taking sides in the debate between Copernican and Ptolemaic astronomy (but certainly in full knowledge of which side his observations would support), Galileo published his findings in 1610 in his book *The Starry Messenger*.

The reaction to this book was vehement and resulted in a number of public accusations of heresy against Galileo, which ultimately led to his appearance before the Roman Inquisition in 1616. It is important to realize that this first appearance by Galileo before the Roman Inquisition was voluntary and resulted from his desire to clear his name from a number of false charges made against him by his enemies. In 1614 a young Dominican priest named Thomas Caccini had given a fiery sermon in Florence in which he had denounced Galileo as a heretic. As a result of the stir that this created, Niccolo Lorini, another priest, copied a letter that Galileo had sent to one of his own disciples, outlining his views, which Lorini then sent to the Roman Inquisition. Later, after another one of Galileo's letters, this time addressed to the Grand Duchess Christina, was heard of in Rome, Galileo decided it was time to appear before the Inquisition to answer the charge of heresy.[1]

In Rome, Galileo appeared for his hearing at the home of Cardinal Bellarmine, an acquaintance of his and a fairly liberal friend of scientific inquiry. Bellarmine's instructions from Pope Paul V were unambiguous: Galileo was to be informed that two of his propositions were censured and that he could no longer hold or defend them. Each concerned the motion of the earth. This prohibition came to be known as the Edict of 1616. If Galileo resisted this order, Bellarmine was to give him a personal order that he could not hold, defend, *or teach* these views. What was clear, however, was that if Galileo did not resist, he was under no personal order any more stringent than that which was to bind all other Catholics at that time, which allowed *debating* Copernicus's theory, as long as one did so in a hypothetical fashion.

What actually occurred at Bellarmine's home is the subject of scholarly dispute. The most reliable accounts indicate that Galileo did not resist this order, indeed that he was given no time to respond to it whatsoever, before another church official took over the proceedings and made a special order, in the name of the pope, that Galileo could not hold, defend, *or teach* either of the two propositions that had been censured. Galileo agreed, and both the substance of the statement and Galileo's assent were duly recorded in the minutes. It is highly likely that Bellarmine then admonished the priest who had gone beyond the wishes of the pope, in that the document testifying to Galileo's "special order" remained unsigned and unenforced.

Later, when rumors began to spread that Galileo had been punished and forced to abjure his belief in the motion of the earth, Galileo turned to Bellarmine for an affidavit that this had not occurred. He secured such a document, which made it expressly clear that Galileo had merely been personally informed of the order that bound all Catholics not to hold or defend beliefs in the motion of the earth, and that no special punishment or orders had been given to Galileo.

Subsequently, Galileo turned his attention to other scientific work, not directly related to Copernicanism, until in 1624, upon the death of Paul V, Urban VIII became the new pope. Galileo visited the new pope, found him quite sympathetic, and dedicated his new book, *The Assayer*, to him. Indeed in six audiences with him, Galileo learned that if it had been up to Urban VIII, the Edict of 1616 would never have been written (although now that it had, he would not rescind it), and Galileo secured permission to publish a new book that he was planning, which concerned the tides. This new book would bear more directly on Copernican theory, albeit, he promised, in a hypothetical way. Thus, reasonably concluding that his hearing in 1616 had not forbade him to treat the subject of Copernican theory as a hypothetical matter, and having secured the permission of the new pope to proceed working on his theory of the tides, Galileo began working on a new book entitled *Dialogue Concerning the Two Chief Systems of the World*.

When the book appeared in 1632, it was a big seller. Within a few months, however, its sale was suspended by the pope, who was very angry with Galileo. It seems that in the meantime, the pope had been shown the unsigned memorandum from the hearing of 1616 and had concluded that Galileo had breached an official papal order. Adding to the pope's pique, in his book Galileo had included one of the pope's favorite arguments against the Copernican system in the mouth of a commentator named "Simplicio." Galileo was ordered to Rome to stand trial before the Roman Inquisition.

When Galileo appeared, he was questioned about the memorandum and replied that he was not aware of any special order that had been given to him in 1616. When the question was pursued, Galileo produced the affidavit given to him by Cardinal Bellarmine (now dead), which spoke clearly to the point that Galileo had not been given any special order not to teach Copernicanism, which his new book would have violated. Having vindicated himself, it was soon realized that Galileo nonetheless could not be acquitted without substantial damage to the Office of the Inquisition. It was privately arranged that Galileo would plead guilty to a somewhat lesser charge of having unknowingly taken his arguments too far, out of "vanity and arrogance," in return for a light sentence. The "plea bargain" was submitted for the approval of the pope, who then decided that the actual sentence was to be an extremely harsh one: Galileo was to be subjected to interrogation under the threat of torture and compelled to abjure on his knees his belief in the motion of the earth, to be followed by lifetime imprisonment and the banishment of his book.[2] Though crushed by his sentence, with no other choice but torture and death, Galileo followed the pope's order. That, upon rising from his knees, he was heard to mutter under his breath, "yet it moves," is probably apocryphal, for this action surely would have resulted in Galileo's immediate execution. It is more likely that this phrase was uttered many years later, when he was safely under house arrest. Thus was born a legend in martyrdom for the freedom of scientific inquiry.[3]

There is perhaps no more compelling story in the history of science that demonstrates the remarkable ideological resistance faced by early scientific inquiry and no finer example of true intellectual courage in all of human history than that of Galileo. To those who would argue that natural science had it easy in comparison to social science, there stands no better refutation than the fate of Galileo. And it should be remembered that Galileo was fighting for the freedom of all scientific inquiry to proceed on an empirical and nonideological basis. When we are able to gather evidence for ourselves, he argued, what is the point of relying so heavily on authority and ideology? Thus may the example of Galileo's courage in the face of the forces of resistance to knowledge give heart to those who wish to pursue a genuine science of human behavior.

But the lesson that social science may learn from Galileo does not end here, for remember that it was also under Galileo that scientific inquiry underwent a methodological revolution, when it split off from philosophy to become an empirical and experimental discipline in its own right. Prior to Galileo, physics was a branch of philosophy and was governed by the rational principles that Aristotle had set down for all of nature. The truths about nature, that is, were to be deduced from a priori principles, and not on the basis of actual observations or experiments. If Aristotle said that heavier objects fell faster than lighter objects, then that was how nature must behave. It did not occur to the "physicists" of the day to go out and see for themselves how nature operated. Indeed, the stunning thing to realize is that to the extent that actual experiments did contradict Aristotelian principles, those principles remained unquestioned.

Although it is hard today to imagine such a state of affairs, it was nonetheless true that before Galileo, "natural philosophy" was beholden to the authority of the divine text of the Bible and the (supposedly) divinely inspired genius of Aristotle. Gaining knowledge from firsthand observations, running controlled experiments, or otherwise testing the authority of long-held propositions was not how knowledge about the world was to be gathered. Before Galileo, scientists had neither the notion that knowledge could be gained through sensory evidence nor the courage to stand up for the truth of what they saw with their own eyes. Science did not yet have the "scientific attitude."

Was it any wonder, then, that Galileo's work was so revolutionary and so threatening to the status quo? Here was a scientist who was skeptical of principles that had not been tested and who felt that reliance on authority—even biblical authority—was superfluous when the truths about nature could be gathered firsthand through sensory evidence. Reliance on reason alone, when uncorrected by the senses, led only to the sterile theories of academic philosophers, who had not the least idea of how nature actually worked. Thus, according to Galileo, science needed to be founded on an empirical basis. And the success of this method sparked a scientific revolution.

What import do these observations have for the social sciences? Plenty. Many would hold that the current state of social scientific inquiry is comparable to the state of physics before Galileo. Often in social science, just like the scholastic philosophers of Galileo's time, we rely inappropriately on authority and ideology to decide empirical questions; we try to solve social scientific problems by relying on our reason and our intuitions, just as Aristotle did in the study of nature. Moreover, in social science, we seem reluctant to test many of our favored propositions about human nature and seem skeptical that we can learn much from social scientific experiments. Social scientists often eschew experimental data (even where they are available) that could guide social policy if they do not jibe with intuition. In short, when dealing with questions concerning human behavior, we practice a methodology of inquiry that seems more like that of Aristotle than the experimental method advocated by Galileo.

Now it is indeed true that in social science, unlike sixteenth-century natural philosophy, we are rarely beholden to the authority of a single theory. There are lots of competing theories in social science. But how do we choose among them? Largely on the basis of how well they match with our intuitions or the fruit of our reason alone. Indeed, it is amazing how many social scientists are disinclined to test a social scientific theory against anything other than their own intuitions. What we need, I contend, is to cultivate an experimental method in social science. We need to be able to weed out the bad theories. We need to learn the lesson in social science that Galileo taught in natural science—that we do not know the truth of any proposition until it has been tested against the evidence.[4]

Of course, it is true that many social scientists extol the virtues of testability; often, however, this amounts to nothing more than lip-service. For how do we run our tests? Often in social science, a hypothesis is tested only by seeing how many confirming instances can be gathered in its support. This, however, is the weakest possible test of the scientific value of a theory, since even if it is successful, it rules out only inconsistency. What is missing is the attempt to falsify the theory, to genuinely test it by specifying beforehand what evidence would refute it, and then looking far and wide for precisely this result. That such a method would eliminate so many currently favored social scientific theories is not a mark against it but a consequence to be embraced.

It is important to realize that it is never a simple matter—in either social science or natural science—to adequately test a theory. Yet often in social science, we do not even try, and hide behind the difficulties of gathering data and testing hypotheses—and the possibility of artificially introducing confuting factors—as an excuse for not making social science more experimental. True, there are difficulties in designing good experimental procedure that any experimental social science will face. But the question we have to ask is whether these difficulties should prevent us from pursuing a science of human behavior. The fact that such barriers are overcome even today in many areas of social science, and indeed that similar barriers are faced in other fields like evolutionary biology, should give us reason to be more optimistic. Instead, the question we should ask is this: Have we done all that we can to make use of the tools that are so far available to us? And we should also ask ourselves whether we are too eager to hold social scientific hypotheses up to a higher standard of proof than is realistic, realizing that natural science also suffers from a common set of barriers to perfect testing. Are we, in short, hiding behind the difficulties of testing social scientific hypotheses because—like the scholastic philosophers in Galileo's age—we are afraid of what we may find out? Are we using the difficulties of experimental social science as an excuse to resist knowledge?

Each individual social scientist must examine his or her own conscience in answering such questions. Yet in doing so, it seems reasonable to keep in mind the human predilection for self-exemption. Do we honestly think that today's social scientists are any less likely to fall prey to resistance to knowledge than the scientists of Galileo's time? Do we think that we are special? Indeed, without the constant threat of refutation by empirical evidence, what would lead any scientist to have confidence that his or her scientific theories were free of ideological bias? We may safely conclude that resistance to knowledge is no less important a barrier to current social science than it was in natural science at the time of Galileo. Indeed, as suggested in chapter 3, in their private reflections, how many of today's social scientists have wondered whether social science is now beholden to the forces of political ideology in the same way that natural science before Galileo was beholden to religious ideology?[5] Though the answers to these questions are bound to be diverse—reflecting differing degrees of skepticism about the ease of employing empirical methods in social inquiry or divergent assessments of the intellectual objectivity of social scientists—is there really any social scientist who can argue that social science should not be responsive to evidence and that what we need is more compliance with the demands of political or religious ideology in our theorizing? Who today would like to be remembered as the Pope Urban VIII of the debate about social science?

In social science we need someone who is willing to stand up for the separateness of social scientific inquiry from religious or political ideology. We need to reaffirm that the proper focus of social science is empirical evidence. Political or religious convictions about gender, race, class, freedom, the existence of God, or human autonomy have no basis whatsoever in deciding social scientific disputes, any more than the Catholic church or Aristotle had a legitimate role in deciding natural scientific ones. To the extent that we are interested in pursuing social science because of its potential role in defining social policy, it is well to remember that if our greatest concern is how to improve the social world, our policies will be most useful only after we have gathered the facts. And to those "sophisticated" critics of the possibility of objectivity in scientific research, may we agree to settle only for as much objectivity as is possible in natural science?

The lessons to be learned by social scientists from the example of Galileo are many. Social science needs both more experimental work and more courage in defending its independence from various ideologies. And in Galileo, we find the embodiment of both of these virtues. It is often said that social science will not come of age until it has "had its Newton." But Newton was a theoretician. And social science has a plethora of theories. Indeed, we are choking on them. We could even have the correct theory of human nature right under our nose, but how would we know it? Indeed, would a Newton of the social sciences today have a chance to get his theory adjudicated against all of the others and to show that it was preferable to its competitors? Are we at the point yet where we would recognize such a theory solely by its accordance with the empirical evidence? Likely not. What we need first is to found social science on an experimental basis, so that we may decide between theories on the basis of their accordance with empirical evidence. We need to be able to weed out all of the bad theories that, in the absence of adequate testing procedures, will continue to find their adherents. Social science is not yet ready for its Newton. What it needs first is its Galileo.

Darwin

If the Galileo affair were the only example in the history of natural science of a clash between ideology and science—of the destructive power of resistance to knowledge—the comparison I hope to draw between the natural and social sciences might be underdeveloped. Yet it is clear that throughout its history (even since Galileo), scientific inquiry has continued to fight against vested interests and the ideologies that support them. Indeed, it is still happening.

Until fairly recently, the clash between religious ideology and Darwinian biology may have seemed remote (though certainly more recent than Galileo's battle with the Roman Inquisition). Most readers are probably familiar with the fact that Darwin's theory of evolution by natural selection faced enormous opposition in its own day (in the late nineteenth century), and even in relatively recent times (in 1925), at the Scopes Monkey Trial.

In that trial, John Scopes, a public school teacher in the state of Tennessee, had been accused of violating a state law prohibiting the teaching of Darwinian biology in the public schools. The fact that Scopes had in fact violated the statute was not in question. Contrary to popular belief, he lost the case, and the statute continued to be carried on the books until it was repealed in 1967. The case attracted enormous popular attention, however, precisely because Scopes had willfully violated a law that he felt stood in the way of honest scientific education and patently violated the constitutional guarantee of separation of church and state. Clearly, the case was about much more than the question of Scopes's guilt or innocence and became a referendum on the freedom of scientific inquiry and education. When the case attracted such a celebrated trial lawyer as

Clarence Darrow to argue in Scopes's defense, and William Jennings Bryan to stand as a witness for the prosecution, the whole country began to choose up sides. Despite the fact that Scopes lost the case, the embarrassing backwardness of the state's attempt to subject scientific education to ideological control won him many supporters, and he became a martyr for scientific freedom. Indeed, the events of the case inspired a play and a popular movie, *Inherit the Wind,* wherein Scopes became a champion of intellectual freedom in American folklore. Thus, just as in the Galileo affair, the victory of the forces of resistance to knowledge proved to be not only temporary but hollow.

For many decades this issue remained out of the public spotlight. In recent years, however, a furious contemporary debate has grown between Darwinian biology and religious ideology that is chilling in its similarity to the Scopes trial over three-quarters of a century ago. Once again, the issue concerns what is appropriately taught in biology classrooms in the public schools. This time, however, the anti-Darwin strategy has evolved: whereas the previous agenda had been to have Darwin banned from the classroom, the argument now is that "creation science" or "intelligent design theory" should be given equal time.

In 1981 the state of Arkansas passed Act 590, which required that public school teachers give "balanced treatment" to "creation science" and "evolution science." It is clear from the act that religious instruction was not to be included in any way as evidence for "creation science," but that it should be limited only to its "scientific evidence," so as not to violate federal law. Indeed, the act went on to state that the current model of teaching only "evolution science" *itself* violated the separation of church and state in that it would tend to provide an environment hostile to "Theistic religions" and would give preference to "Theological Liberalism, Humanism, Nontheistic religions, and Atheism in that those religious faiths generally include religious belief in evolution."[6] Clearly rooted in the belief that evolution by natural selection was just one theory among many (and indeed was itself the basis for a religious ideology), the effect of the act was to put forth "creation science" as a legitimate scientific contender to Darwin's theory of evolution.

It is my hope that the ignorance and scientific illiteracy of Act 590 is so apparent that it need not be dissected here point by point. Just as I take it that in the example on Galileo I did not have to convince my readers that Galileo was right that the earth does indeed move, I take it that anyone who has made it this far into the book does not need to have explained to them the overwhelming scientific evidence for Darwin's theory, or the intellectual bankruptcy of "creation science." Suffice it to say that there are several excellent books now available that demonstrate decisively that "creation science" does not meet any reasonable definition of "science," is in fact motivated solely by religious faith, has no genuinely "scientific" evidence to speak of in its favor, and is cloaked in the mantle of science only to lend it an air of respectability in the fight against Darwin.[7]

Happily, I can also point out that the constitutionality of Act 590 was successfully challenged in court, with some of the most respected scientific and philosophical authorities in the world lending their support to overturn it. No mere moral victory this time, science won a decisive battle in its struggle with the forces of resistance to knowledge. The ruling found, among other things, that it was ludicrous to claim that Darwinian biology could be put on par with a secular religion, that the defense could produce no evidence for the truth of creationism that went beyond fruitless attempts to discredit evolution or reliance on the Bible as an authority on natural events, and, most telling for our current focus, that "the methodology employed by creationists is another factor which is indicative that their work is not science. A scientific theory must be tentative and always subject to revision or abandonment in light of facts that are inconsistent with, or falsify, the theory. A theory that is by its own terms dogmatic, absolutist and never subject to revision is not a scientific theory."[8] Thus is the current attempt to have "creation science" given equal time with evolution by natural selection in our nation's classrooms revealed as nothing more than an intellectual fraud, motivated by deepest grounding in the forces of resistance to knowledge.

Despite the decisiveness of this victory, however, the fact that such efforts were even necessary cannot help but sadden those who care about the independence of science. The parallels between the ideological reaction against Galileo and that faced by Darwin, coming centuries later, are depressingly similar. Sadly, it is also my duty to report here that the "creation science" lobby has now regrouped and is continuing to push its agenda in school districts across the nation. In 1999, the Kansas Board of Education voted to delete the teaching of evolution from the state's science curriculum (although in 2001 Kansas educators voted to restore it). In 2005, however, a newer and more conservative board decided to add "intelligent design" to the science curriculum in order to achieve greater "balance." In 2002, the Cobb County (Georgia) School Board mandated that stickers be put on all high school biology textbooks that read, in part, "evolution is a theory, not a fact"; in 2005, a federal judge ruled that the stickers were unconstitutional and ordered them removed. In Ohio in 2002, some members of the state school board sought to include the teaching of "ID theory" (intelligent design theory) as a mandated companion to evolution in the biology classroom. In Pennsylvania in 2004, the Dover Area School District voted to mandate the teaching of "intelligent design" as part of its science curriculum. The voters spoke by promptly turning the school board out of office at the next election and the ACLU filed a federal lawsuit, which it won. In a strongly worded opinion that was remarkable in its similarity to the reasoning used to reject Act 590, the judge concluded that intelligent design theory was "not science" and therefore had no place in the science classroom. He went on to cite the "breathtaking inanity" and "striking ignorance" of the school board's original actions. Such a clear victory in this specific case is nonetheless rendered poignant by the fact that 25 years after Act 590, essentially the same argument had to be made again, and it remains in no way clear that similar success can be anticipated against the numerous efforts that currently are being made to do in other states what could not be done in Pennsylvania.

Won't people ever learn? Must science forever fight the same battles? And if natural science, with its enormous track record of success, cannot overcome resistance to knowledge once and for all, what hope have we for the social sciences? Should we, in keeping with T. H. Huxley's observation that "life is too short to occupy oneself with slaying the slain more than once," just declare the battle won and go back to our laboratories?

No, for this is the wrong way to think about resistance to knowledge. Resistance to knowledge is a living phenomenon, and it is prepared to rear its head at every turn where the results of our empirical inquiry do not meet with the reigning religious or political pieties. It is a myth that science will ever be done in its battle against resistance to knowledge—that it will ever finally "win"—just as it is a myth that scientific inquiry will ever come to an end once we have found out all that there is to know. The growth of scientific knowledge is open-ended, and science (whether natural or social) must constantly prepare to do battle with the forces of resistance to knowledge anew with each controversial discovery.

Moreover, scientists should be aware that the guises of resistance to knowledge are many. Throughout its history, natural science has faced opposition primarily from religious ideology. But may it not also come, as it has in the past, in the form of governmental ideologies, such as the Lysenko affair in Soviet genetics, or the cruel and pointless human twin experiments under the Nazi regime? May it not take the form of the soft-sounding admonitions of political correctness that behavioral geneticists ought not to investigate anything other than environmental causes of group differences for important human traits, for fear of what we may turn up? May it not arrive in the human desire for connectedness with one another and with our environment promised by today's new age movements of crystal healing, ecofeminism, aromatherapy, and alternative medicines, which are behind the current movement of "antiscience"? And may not all of these sources, and others, pitch themselves even harder against the empirical findings of social science?

Indeed, some would say that politically motivated resistance to scientific knowledge is as close as today's White House and that in this battle, even the natural sciences are losing. The decision by officials at the Environmental Protection Agency to dismiss as "unproven" the consensus of worldwide scientific opinion on the subject of global warming—culminating in the 2001 decision of the United States to pull out of the Kyoto Protocol (which mandates a worldwide effort by signatory nations to reduce greenhouse gases by 2012)—is perhaps the best recent example of resistance to knowledge influenced by political ideology. As a coda, one might consider the Bush administration's decision to severely limit the number of stem cell lines that are open to researchers who receive federal funding.

Indeed, things have gotten so bad recently in the relationship between scientists and government officials that in 2004, the Union of Concerned Scientists took out a full-page ad in *Time* magazine, that read in part:

> We all trust our government to use facts when making decisions about things like mercury pollution, leading poisoning, and breast cancer. But the evidence shows that officials at the Environmental Protection Agency, Health and Human Services, and other federal agencies are trying to spin, manipulate, and even hide scientific facts to further a political agenda. That's why more than 5,000 of America's leading scientists have called on our leaders to stop distorting scientific knowledge and restore scientific integrity in government. The scientists include: 48 Nobel laureates, 62 National Medal of Science recipients, 129 members of the National Academy of Sciences.[9]

The ad was intended to call attention to the release of a report entitled "Restoring Scientific Integrity in Policy Making" that detailed numerous instances of "manipulation of the process through which science enters into . . . decisions" made by the Bush administration.

Such struggles between science and the forces of resistance to knowledge represent an eternal tension: that between what we have reason to believe and what we would like to believe. Understood in this way, it is clear why the method of scientific inquiry, with its insistence on empirical evidence, guarantees an ongoing clash with ideology. Thus, it should not surprise us that social science faces these same pressures in its struggle to become a genuinely scientific enterprise. Natural and social science both must face resistance to knowledge in their quest to base our beliefs on a rational consideration of the evidence, and not on the seductive calling of our ideologies. Thus in science, too, the price of freedom is eternal vigilance.

Lessons for the Social Sciences

I have already drawn some conclusions for the social sciences along the way in this chapter. But it is now time to consider explicitly what the social sciences may learn from the historical struggles that natural science has had with resistance to knowledge. In what way is the situation now facing social science similar to or different from that faced by natural science? What have we learned about our conception of science that both natural and social science may hold in common?

First, it is important to meditate on the question of what is distinctive about science. Regrettably, many who have envied the success of natural science have been in a hurry to frame their conclusions and have supposed that there is some sort of "scientific method" that can be followed by other disciplines, in their attempt to achieve scientific rigor. One often hears that what is needed in the social sciences is for them to (1) observe, (2) frame a hypothesis, (3) derive a prediction from it, (4) test that prediction, and then (5) revise the original hypothesis on the basis of the evidence. This is the classic "five-step method." Unfortunately, throughout history, several of the social sciences have attempted a slavish imitation of this methodology without really understanding the problem with it—that there can be no recipe for growing science—and have succeeded not in achieving truly scientific results, but only in practicing what is known as "scientism."[10] Indeed, it has long been recognized even in the natural sciences that the so-called scientific method is a fiction. For how does one start with observations without having some sort of preliminary hypothesis? How do we know what to observe? And how do we know on what basis to revise our hypotheses? Among those who analyze scientific reasoning, these and other problems have led to widespread repudiation of the idea that there is such a thing as the scientific method behind the natural sciences.

Yet this dispute distracts us from an important issue. For perhaps there is something distinctive about scientific reasoning that, if we are patient enough to study it, may inform our attempt to revolutionize the social sciences. Indeed, despite the problems presented by the search for a simple scientific method, it is perhaps nonetheless true that what is special about science is its methodology. Thus, it does seem appropriate at this point to ask what is essential to scientific reasoning and how natural science has achieved its amazing discoveries. Let us start by examining some insights that have been buried in the controversy over scientific method.

Despite its obvious oversimplification of the complex and diverse procedures by which the natural sciences work, there are some important lessons about scientific reasoning that can be learned by examining the spirit of the five-step method. First among these is that scientists should care about empirical evidence. While it is true that an array of difficulties must be faced in trying to codify exactly how evidence should be used in individual cases of scientific reasoning, it is nonetheless important to realize that one defining characteristic of science is that if one is going to do science, some attempt must be made to reconcile one's theory with the facts that have been gathered on the basis of our observations. Whether we have gathered that evidence by telescope, microscope, or simple observation (though it must be sensory evidence, albeit sometimes tempered by mathematical insight), it should count in assessing the worth of our hypothesis. Whether we formulate our hypothesis first or last, what matters is that as Galileo taught us, empirical evidence is central to the job of testing scientific theories. Indeed, without it, can one honestly say that we are doing science? The nonempirical theories of fifteenth-century Scholastic philosophers, who were content to offer their hypotheses only on the basis of their fit with intuition, or the latter-day "creation scientists," who formulate their hypotheses on the basis of their conformity with the written word of the Bible, *these are not science.* Why not? For the simple reason that science cares only about whether our beliefs fit with sensory evidence, and not about what the consequences might be for our other beliefs.

How then is empirical evidence used? This brings us to the second essential feature in scientific reasoning. In science, evidence is used not only to formulate but to test our hypotheses and theories about how the world, and even our own behavior, might work. As previously noted, the use of evidence made by science is primarily negative. We are looking to see if our hypothesis is consistent with the evidence; if it is not, then no matter how beautiful or plausible or congenial it may be to our system of beliefs, we must, if we are to remain scientists, modify it or dispose of it altogether. Of course, it is hardly ever a straightforward matter for a scientist to know whether a falsified prediction is the result of a bad theory, a false auxiliary assumption, a faulty piece of apparatus, or something else. And in the meantime, while evidence is still being gathered, it may be that a theory is retained despite some lack of fit with the evidence. Indeed, as recent philosophers of science have pointed out, all scientific theories have such problems, and the process of scientific practice is complex enough that it can sometimes take hundreds of years to come to an adequate conclusion.[11] Science is in a perpetual search for better and better theories. But the important thing to remember when presented with all of the messy details about how scientists test their theories is what is behind it. And this is respect for the power of evidence sufficient to overturn even our most favored theories once we learn that they are inadequate. It is, in short, possession of the scientific attitude toward evidence, detailed in chapter 2, that rests on the conviction to believe something only on the basis of its empirical evidence, and not to be biased by what we merely hope to be true.

Even if there is no simple scientific method that we may follow in our attempt to make the social sciences more rigorous, we may yet learn something from this scientific attitude toward evidence that is present in the natural sciences. It is essential to our progress in attempting to understand the kind of social problems we face today, and to do something about them, that we embrace an attitude toward evidence that ensures that we are not making the same mistake as the medieval Scholastics or the "scientific creationists." If we wish to do something about the horrible state of the world today, we cannot afford to indulge our intuitions while we let the gathering of empirical evidence languish. If we are serious about social reform, aren't we better off making sure that we emulate a methodology that gives us as little risk as possible that we are "fooling ourselves" about the problems we are trying to solve? And isn't it appropriate that if social science is going to serve as the basis for our social policy, it should be based on the best empirical evidence that we can gather? The answer to both of these questions is a resounding yes! Indeed, in light of the importance of our task, we should not be at all timid in proclaiming that both the study of nature and the study of society are engaged in a similarly "scientific" enterprise.

This last point is often missed as a result of the huge gulf between the success enjoyed by natural science and by social science. It is probably easy for a contemporary social scientist to look at today's natural science and see no real basis for comparison. Natural science today, bolstered by the technological fruit it has born, is a success story of unparalleled proportions. Thus, it is understandable that many social scientists look in awe at the achievements of natural science and are reluctant to draw any comparisons.

Yet what is often missing in such thinking is a kind of intellectual empathy for those natural scientists who were working before such runaway success had even been imagined. In hindsight, it is easy to underestimate their situation and to see the outcome as inevitable. Yet why, then, did the basic theory of the motion of the earth, first proposed by Aristarchus in the third century B.C., and later codified by Copernicus in the sixteenth century A.D., not become accepted even by scientists until many centuries later, after Galileo's imprisonment? What must it have been like to live then, and to be in such a minority against the forces of resistance to knowledge, and yet to know that you are right. It is time to see the truth: that natural science succeeded not because the truths that it found were inevitable, or because it did not face resistance to knowledge, or because the natural world is so simple. Likewise, it did not succeed merely on the strength of a powerful methodology. A fair reading of the history of natural science leads one to the conclusion that an indispensable factor in the success of natural science was the intellectual courage of its practitioners.

The lesson that the social sciences may learn from the natural sciences, therefore, is not just methodological; it is practical. It is a lesson in the power of persistence and discipline, and the virtue of sticking by the evidence. Science is not a magic process. It is not even self-correcting. It works because of the diligence of scientists in following the dictates of the scientific attitude, in ferreting out weak hypotheses by comparing them to the evidence. It works because of a willingness to hear what the evidence tells you, even when it clashes with the reigning ideology. It works because scientists know that their work will be held up to the highest standards of scrutiny by their peers, whose competitive spirit motivates them to publicize any errors. *This is precisely what is needed today in the social sciences.*

Natural science overcame resistance to knowledge and developed a new methodology of inquiry that resulted in the scientific revolution of the seventeenth century. May the social sciences now follow this same path?

Conclusion

I hope it is clear what the history of natural science may teach us about the prospects for developing a science of human behavior. But it is now time to answer a question that has probably already been framed in all readers' minds by now: Is it reasonable to suppose that the social sciences can actually do it? Is there anything disanalogous about the comparison between natural and social science that should give us pause before we proceed in our task? If not, how are we to explain the enormous success of the natural sciences while the social sciences (which are at least as old) have languished? Aren't there some respects in which the situation faced by social science is different from that faced by natural science? What does the "scorecard" look like so far?

Let us start with some reasons for optimism:

1. Arguably, both natural and social science face the same sort of barriers. Complexity, open systems, subjectivity, and resistance to knowledge are present in both types of inquiry.
2. To the extent that their tasks are similar, social science has the benefit of a forerunner to show the way. Natural science has provided social science with both an example of a courageous fight against resistance to knowledge and an effective methodology of inquiry.
3. As in natural science, the issues faced in the study of society are empirical. And there is no better method than science for studying empirical issues.

Let us now turn to some reasons for concern:

1. Natural science has the usefulness and availability of technology in its favor. It is a practical success. This practical success compensates for the loss of our favored ideologies. The practical benefits of social science have been slower to develop and are more difficult to appreciate.
2. There is perhaps less prejudice about the way nature works than about the matters that social science confronts. The way the world is and our relationship to it, though especially important in ancient times, is still not nearly so controversial as the issue of whether we are in control of our own behavior.

What are we to make of the latter set of issues? First, while it is indeed true that there are as yet few tangible practical benefits (at least popularly known) to the social sciences, there is no reason to think that these would never become available. Indeed, remember that the whole point of a science of human behavior is to improve the quality of human life. And it is also important to point out that natural science too was slow to develop these fruits. Should we never attempt to understand the causal forces behind human behavior because we, at first, may not know how best to make use of them?

Second, one should beware of underestimating the power of ideologies just because, to us, they seem remote. While it is true that the medievals would have been extremely upset by the suggestion that we can have a science of human behavior, the proposition that we could have a science of nature was no trivial matter to them. The Aristotelian worldview, which put earth at the center of God's creation and made the heavens perfect, or the pre-Darwinian worldview, which held that humans were at the top of the animal kingdom, were deeply important facets of the Christian faith. Recall that the weight of these issues was such that until relatively recently in human history, those who questioned them were often put to death.

Nonetheless, it is probably fair to say that in the great hierarchy of egotistical beliefs that have made up human ideologies over the centuries, the proposition that humans do not have a very good understanding of why they behave as they do (and that we can gain a better understanding of this through science) is the most threatening rebuke to our conception of ourselves. Until fairly recently, few had the courage even to pose the question. The job facing a true science of human behavior will not be easy, and there is no sense in underestimating the task that lies before us. Perhaps the barriers are greater here than they were in the founding of the natural sciences. And we must admit squarely that as of yet, we have few ready practical fruits to smooth the way. Indeed, put this way, it is no wonder that the idea of a science of human behavior is resisted by most people, for all it offers is the bitter pill of self-examination with no compensations or illusions to soften the blow.

But we must have courage in our fight for a science of human behavior, for it is our best hope for salvation from the host of social miseries that afflict us. We must resolve not to listen to the smug warnings of those who, like the opponents of Galileo and Darwin, are afraid of what we may find out. Indeed, we must be willing to make the same effort on behalf of social science as our scientific ancestors did on behalf of natural science. Like them, we must resolve to disregard the pieties of the day when we are engaging in scientific inquiry. For if we are to discover the truth about ourselves, it would surely transcend the prejudices of any one generation. Of course, it will not be easy. Yet the world awaits the efforts of social scientists who are willing to improve the human condition. Our social problems will continue to consume us in their misery until we can find a solution. And in the long run, the triumph of science over this last domain of human resistance to exact inquiry seems inevitable. The facts about our behavior are empirical, and science is well equipped to study such relationships.

We are, after all, merely one species, inhabiting the third planet of an unremarkable star, tucked away in the corner of the universe. As a species, we have not even been here that long. May we not therefore comprehend the enormity of our ego in claiming that there can be no science of our behavior, in contradistinction to every other matter in the universe? And does not a better future await our willingness to apply to our human problems every ounce of our reason and intellect? Indeed, if we are so special, surely what is special about us is our brains. Let us therefore use them to their utmost in elevating the way that we live and treat one another. How will our efforts be judged by future generations? If we fail even to try to improve our lot, do we not deserve to be thought of as existing in a Dark Age?

Resistance to knowledge must be overcome. We must be ready to face the truth about ourselves if we are ever to improve our social relations. Indeed, what greater challenge could fall to social science than this? For anyone who has ever lamented living in the modern era, when there is so little truth left to discover (and who hasn't thought that *they* could have been a great scientist if they had just lived 500 years ago?), there is perhaps no more exciting time to be a social scientist. Who can know to whom the defining moment in social science will fall, as in natural science it fell to Galileo, when social inquiry at last steps out from the shadow of natural science and proclaims that it too is a science. And, more important, who can foresee the potential benefits for the future of the human race of finally approaching such horrifying social problems as child abuse, for instance, not just on the basis of our intuitions and our emotions, but to study them with all of the powers at our disposal, so that we shall know when we look into a child's eyes that, just as science has eradicated smallpox, there are other sicknesses that a child need never fear when at last we learn their true cause.

5 What Is to Be Done?

Philosophers have only sought to interpret the world. The point, however, is to change it.
—Karl Marx

Having now explored most of the arguments necessary to establish that we both need and can have a science of human behavior, I would like to end this book with a parable. Imagine arriving on a planet bare of any technology, with only minerals, vegetable matter, and animal life. How would you survive? What would you be able to do with what you had? Now consider that this was exactly what earth was like when humans first evolved into a separate species on this planet. All that has been built, which we now use and take for granted in our everyday life—fiber optics, digital computers, plane travel, nuclear weapons—has been manufactured as a result of human ingenuity in using the raw materials found on earth when we arrived here. Constrained only by the laws of physics and the limits of our imagination, we have built a technological society that seems unimaginable given our humble beginnings.

Now reflect for a moment on the type of human being we find inhabiting this technological utopia. As did our primitive ancestors who lived before the age of science, we fight wars, we lock up resources in the hands of the few while others go hungry, we kill for spite, we ignore the suffering of children, and we hate one another based on superficial differences in our physiognomy. Except for a few shining examples of human invention of beneficial social arrangements (like democracy or civil rights), aren't most of our social arrangements—our animosities, our response to danger, our forms of courtship—pretty much the same as they have been for hundreds of thousands of years? And yet how little do we understand them? And how little have we done to improve our way of life? But for the trappings of our technology, most of us live a life nearly indistinguishable from that of our distant ancestors. Can't we do better than this? Couldn't we apply the same amount of ingenuity to improving our society as we have to the development of technology? Indeed why not apply the highest form of human reason, science, to our social problems? Why can't we attempt to find better solutions to the problems of war, crime, racial strife, and child exploitation?

The picture I have just sketched is familiar to most of us; it is often referred to as the human condition. But need it be our permanent state of existence? Is there anything we can do to make our lives better in the way that we interact with one another? Yes! We can follow the same path we used to liberate us from our once-meager material existence to improve our social existence. And in doing so, we must rely on the only real resource that justifies our feeling of superiority over other animals and has allowed us to advance far beyond our original circumstances: our ability to reason.

The only way out of our current social dilemma is to apply our reason to the solution of our social problems. We must attempt to gain more accurate knowledge of why we act as we do, so that we may then use our ingenuity to think of better ways for us to live. In doing so, we must be prepared to be honest about what causal factors are behind our behavior, before we can hope to do anything about it. That is, just as we needed to understand the laws of nature before we could harness their power and develop technology, we must now resolve to understand human nature without the veil of ego and lies that we use to comfort ourselves, so that we can build a better society, with less misery, suffering, and cruelty toward one another. This goal can be achieved, I submit, only if we are willing to have a science of human action—only if we are willing to shed the egotistical belief that we already know the basis for our behavior.

To many living in these Dark Ages, this will seem like a utopian dream. But I ask you to imagine what those living in medieval times might have thought of our modern technological order. Besides, what is wrong with envisioning a better social order—even one that may, to us, now seem like a utopia? Why is it that we find it so easy to imagine a future filled with enormous growth in our technological achievements—new gadgets and toys, or even revolutionary scientific breakthroughs in transportation, communication, or agriculture—yet we find it so hard to imagine a future in which we treat one another better than we do today? Why not let our vision of the future—both technologically and socially—be framed by science?

Science works. It is the greatest invention of the human mind in the history of our existence on this planet. It is also the single greatest tool we have to improve the state of our existence. But we must understand it for what it is. Science is not inevitable. Neither is it a foolproof process that will always lead us to the truth. Indeed, science is fragile. In all of my years reflecting on the nature of scientific thought—in light of factors like resistance to knowledge—the amazing thing to me is that science ever happened at all. For science must be nurtured by our refusal to take the short-cuts of ideology and superstition in trying to understand some new fact about nature or about ourselves. To see it flourish, we must have the courage to let it make our old beliefs seem uncomfortable to us, in the face of our new-found knowledge about how things actually work. It requires, in short, that we be willing to face down resistance to knowledge in all of its forms and to challenge our prejudices about free will, human predictability, religion, and political ideology. And once we have done this, may we not then proceed to build a better society based on our deeper understanding of ourselves?

At this late juncture, however, I should take a few moments to address the concerns of those who fear that a science of human behavior may not be used for good but rather for evil—that just as natural science led to modern medicine and space travel, it also led to the atomic bomb. Science, after all, does not always lead to benefits. It depends on how we use it. Indeed, a good deal of evil has come wrapped in scientific pretensions. Marx's allegedly scientific understanding of human nature was responsible, through Lenin and Stalin, for a great deal of tyranny and suffering. And even the Holocaust itself was based on Hitler's bogus scientific insights into racial differences. People are understandably nervous about those who claim to have a scientific understanding of human nature and who seek to use it to control our destiny, for such grand ambitions often end up in fascism. In the wrong hands, might not the effort to reform our society on the basis of a scientific understanding of human behavior lead to despotism and misery? As we consider committing a social science, we should never forget that some of the worst horrors human beings have perpetrated against one another have been justified by an allegedly scientific understanding of human nature.

It is important here to point out, however, that the greatest safeguard against such pseudoscientific misunderstanding of human nature is to pursue a genuine science of our behavior; the best way to combat false belief is not through ignorance, but through knowledge. Those who seek to misuse science in the service of their ideological agendas can best be defeated by making sure that we understand the true nature of scientific reasoning. We may then stand ready to unmask pretenders to the name of science because we have seen it in its genuine form.

How is the kind of scientific investigation I am advocating distinct from those others, both benign and evil, that have misfired in their attempt to understand human behavior? The science I am proposing is one modeled on the central insight that may be drawn from the history of natural science and that seems best able to explain its success relative to other methodologies: that what is distinctive about scientific reasoning is not a set of rules or procedures or even a method, but an attitude that one takes toward the revision of a theory in the light of empirical evidence. This attitude is based on the ideal that science, insofar as possible, must be nonideological; science must have no creeds. Indeed at base, this is what was wrong with Marxist theory, and all of the others whose insights amounted to no more than ideological pseudoscience. One cannot properly use science to back up only those theories that suit our ideology or our intuition. Ideologically driven social science is a perversion of the scientific attitude. Rather, a truly scientific understanding of human behavior would allow our understanding of ourselves to be shaped by empirical inquiry. Instead of torturing the evidence to fit our theory, or selectively highlighting only those results that confirm it—as one can plainly see in the works of Marx—a true science should allow our theory to be shaped by empirical investigation. To do otherwise is to pursue only scientism, and therefore to run the risk that even the most scientific-sounding investigations will degenerate into mere pseudoscience.

What then of the fear that the pursuit of a science of human behavior would inevitably lead to tyranny? One may now see that, properly pursued, the scientific investigation of human behavior would not necessarily lead to tyranny but could in fact be quite liberating. There would be power in our knowledge of ourselves that could be shared by all who knew it. Indeed, the way to avoid tyranny by those who claim to know the secrets of our true nature is for scientific knowledge about our behavior to be widely disseminated. Remaining ignorant of the true causal forces behind human action is not going to save us from ourselves. The more we know and are in control of our own destiny through increased understanding of human motives, the safer we are from those autocratic regimes that purport to reveal some secret truth about human nature.

Moreover, I think it is here instructive to look, once again, at the state of the natural sciences, in order to address our concerns over the misuse of science. I think it is not entirely naive to argue that on the whole, despite episodes of misuse, science has been more of a benefit to the human race than a detriment. And despite the risk that we may someday revoke this endorsement, I also believe that if one is advocating an end or a limit to science, one should be prepared to face directly the devastating human consequences of deliberately remaining ignorant. There are real causal forces behind both the laws of nature and our own actions, which govern us whether we are willing to admit them or not. And in light of the horrible human suffering we currently face as a result of disease, starvation, overpopulation, war, crime, and poverty, can we really sanction the prevention of scientific inquiry for fear that it may lead to some future horror? Isn't it better to know the world around us than to remain ignorant of it merely because we trust ourselves so little? Indeed, could we even stop our efforts to know ourselves or the world around us if we wanted to? And if we are going to engage in inquiry, why deliberately stand in the way of the most successful methodology that the human mind has ever invented for gaining knowledge?

It is at this point that I think we must be willing to bet on science, for the litany of social problems outlined in chapter 1 will not go away by themselves, and the enormous human suffering that results from them should not be underestimated. Yes, it is true that we may fail. But we are failing anyway. And I believe the lesson to be learned from the natural sciences is not that science is perfect, but—in concert with earlier admonitions about the dangers of resistance to knowledge—that we have more to fear from ignorance than from knowledge, no matter how unsettling the results of our inquiry may be.

In saying this, I do not mean to prejudge what science will turn up; I am merely saying that we need to be prepared to face the truth even if it is unpleasant. We must cultivate the scientific attitude, and be as curious about human behavior as we might be about that of some new species. We must prepare ourselves for the greatest intellectual challenge that has ever faced humankind: to engage in a truly scientific study of our own behavior.

In the bulk of this book I have pursued a line of argument meant to convince readers of the need for a science of human behavior. But I have said little so far about the practical problems of how to build such a science. How do we get there from here? What work lies ahead? My hope is that in the foregoing chapters, I succeeded in demonstrating that empirical social science is our best bet for remedying the problems that have afflicted our social world. Now, however, it is time to address the issue of what steps must be taken if we are to achieve such a goal. At the beginning of this book, I diagnosed what I thought was wrong with the social sciences. Now it is time to offer a prescription for what we might do to fix them.

First and foremost, we must be aware of the dangers of resistance to knowledge; we must be true to the scientific attitude, which tells us that there are no taboos in scientific research. Those who do social scientific inquiry must have courage in the face of entrenched religious and political interests, especially their own. We must resolve to decide empirical matters on empirical grounds. At all costs, we must beware of ideology. Science has no heretics and cannot function properly if we are afraid to investigate controversial theories merely because they clash with what we hope may be true.

As a corollary, those of us who have already seen resistance to knowledge working against science in other venues should reflect on our motives and intuitions if we still find ourselves resisting the idea that there can be a science of human behavior. Are we being honest with ourselves? In pursuing the line of argument I have outlined in this book, I have often lamented that so many thinkers who so clearly see resistance to knowledge in the form of religious ideology working against natural science (for example in the creationists' prejudices against evolutionary biology) are nonetheless blind to the other forms of resistance to knowledge that have been used to smother a science of human action. Can't they see the parallels? Aren't the egotistical beliefs that we hold about human uniqueness or some of the political beliefs that we hold about race, gender, and class preventing us from pursuing the scientific study of human society? Aren't they too a form of resistance to knowledge? To those who see in creationism a virulent enemy of free scientific inquiry, and yet deny that we can have a science of human action because it offends certain political beliefs, I must question their intellectual honesty. Is their stand against resistance to knowledge one of principle, or is it just that it matches their prejudice against certain religious beliefs? We should be fighting resistance to knowledge in all of its forms, especially when we find ourselves the most certain that we ourselves do not suffer from it. Why not extend our recognition of the chilling power of resistance to knowledge to social science and see how many of our own prejudices have held back scientific advancement in the study of our social relations?

Second, we must draw strength from the analogy with natural science. The objections to a science of human behavior are almost identical to the historical objections that were made to the scientific study of nature. Indeed, resistance to knowledge is the single biggest barrier to a science of human behavior, just as it was to a science of nature. Yet as in natural science, social science can make progress. We should learn the history of the natural sciences and be inspired by their struggle.

Third, we must admit that the questions we have about human behavior have right answers to them and that these answers can be discovered through empirical inquiry. There are genuine causal forces at work behind the problems of serial killing, child abuse, hyperflation, and recessions, but we do not automatically know what these are simply in virtue of being human. We need to engage in the rigorous and systematic study of these problems before we can understand them. We must be prepared to be surprised and disturbed by what we may find out.

In doing so we should pursue a methodology of inquiry that will allow us to learn from our data. Social scientists should use experimentation wherever it is possible. And we must stop searching only for those positive instances that will back up our favored theories instead of the falsifying instances that might refute them. If we are interested only in performing ideological social science, then the search for positive instances is to be expected, for all we are interested in doing is backing up our preferred theories. But if we expect to learn from our data, we must allow scientific evidence to shape our generation of and choice between social scientific theories, such that we are not afraid to abandon them when they do not fit the facts. We must be ruthless in deciding between social scientific theories only on the basis of their empirical evidence.

Fourth, as a consequence of the above, we should not expect the truth about human behavior to be congenial. There is an ocean of truth about the causes of human behavior that is still waiting to be discovered. But all of the happy truths have probably already been found. We must therefore be ready to delve into our fears about ourselves before we may hope to learn the true causes of our behavior. Again, I am not suggesting that this in and of itself has empirical weight, but simply that one should not be afraid to make bold hypotheses, which later can be shaped by scientific testing.

Fifth, we must not listen to the naysayers, who will make sophisticated-sounding arguments about how science cannot be objective, that we cannot discover anything that we do not already know, or that a science of human behavior would be too dangerous. It should be remembered that any scientific revolution faces a tremendous amount of initial resistance by those who seek to protect the status quo. It should also be comforting to know that many of the clever-sounding philosophical arguments against a science of human action work equally well to show why we should not be able to have a successful natural science. Not only can such charges be answered on philosophical grounds, but our efforts are vindicated by the practical success of natural scientific inquiry.

Finally, we must never lose sight of our ultimate goal: to discover the causal factors behind human action so that we can improve the social world. This is not to say that social scientists should not attempt to be objective. The fact that in social science we already know how we would like to use our knowledge before we have gathered it does not mean that our inquiry cannot embrace the objective ideal. Objectivity is a matter of the attitude that one takes toward the refutatory power of evidence. Thus, perhaps social science is less like physics than it is like medicine, where we already have a vested interest in making use of our knowledge—and know what result we would like to bring about—and yet we realize that it serves no purpose to fool ourselves about the correct answers to our empirical questions. In medicine we understand that if we are actually to save lives, we must not allow our hopes and fears to cloud our inquiry into how the course of disease actually works in the human body; our goal is too important to allow ourselves to be distracted from the pursuit of science. Is our goal any less urgent in the social sciences?

Thus, my prescription for the social sciences is largely conceptual, for there is no scientific method that can be followed, or any other recipe by which one might reliably "grow" a new science. Fortunately, there does not have to be. The history of natural science has taught us that the most important factor in determining the success of scientific inquiry has been acceptance of the scientific attitude by its practitioners. Science is a self-conscious enterprise; we need to believe in science before we can do it. And it is the scientific attitude, I maintain, that has been most sorely missing in contemporary social science. It is one thing to give lip-service to the value of empirical inquiry; it is another to believe in the scientific attitude toward evidence so strongly that one grows intolerant of ideological bias, and the insidious creep of resistance to knowledge, when considering empirical matters. Thus, I maintain that just as the barriers to a true science of human behavior have been mostly conceptual, so its course for success can be determined not by factors such as the alleged complexity, openness, or subjectivity of its subject matter, but instead by the determination of its practitioners to respond scientifically to the challenges that are presented by the vicissitudes of social inquiry. In short, the prescription for the success of a true science of human behavior is to start with a shift in attitude away from ideology and toward the refutatory power of empirical evidence in social inquiry parallel to that ushered in by Galileo during the scientific revolution of the natural sciences. Only in this way may we fulfill the dream of understanding human behavior well enough to address our social ills.

If we succeed in reforming the social sciences according to the platform that has just been indicated, what might the resulting society look like? Let us pause here to imagine it. Imagine a society in which the economy was understood well enough that it did not suffer from the yawing ups and downs that bring about periods of poverty despite an abundance of world resources, wherein we are faced with such paradoxes as that we grow enough food to feed the planet and yet there is starvation. Imagine a society in which crime had all but been eliminated, because we understood the causal factors and environmental conditions that led to it, and could intervene early in the lives of those who would, in other circumstances, have become criminals. Imagine a world in which nations worked out their differences through treaties and negotiations because at last we understood the psychological roots of war, and so had gained control over aggressive nationalism by understanding it on a psychological basis. Imagine, in short, the world that most of us have always dreamed of inhabiting and that might be available to us but for our own ignorance and ineptitude in building it.

Is the world that I have just described an impossible ideal? Many will say that it is. But we will know for sure only if we try to achieve it. And, if it is possible, who would like to be responsible for delaying the day when it might arrive? Indeed, who among us feels confident that without radical intervention, we will ever be able to build such a society at all? And who genuinely can doubt that if we do someday succeed in building such an ideal society, the path that will lead us there will be lit by science?

How far are we from the day when we might actually realize some of the goals here imagined? Perhaps closer than one might think. For even now, there are suggestive examples of successful public policy that seems to have been informed by empirical social research. At the beginning of this book, I detailed a litany of the failures of social science adequately to inform our understanding of, and therefore our control over, social events. I would now like to end on a more hopeful note with a true story about a stunning reduction in the murder rate in New York City.

In January 1994 the New York City Police Department instituted a radical policy change in law enforcement. After years of ignoring such "petty crimes" as turnstile hopping, public urination, aggressive panhandling, graffiti, vandalism, and "squeegee operators," it was decided that such offenses were serious problems that adversely affected the quality of life of all New Yorkers. Indeed, the tolerance of such crimes was thought to create an atmosphere of disorder within which more serious crimes might flourish. Thus, the NYPD set out on a vigorous "Quality of Life" campaign, paying attention to those offenses that had a deleterious effect on public order, in the hope that this would deter more serious crime as well.

Support for this policy change was self-consciously based on James Q. Wilson and George L. Kelling's article "Broken Windows," published in 1982 in the *Atlantic Monthly*. In this article, Wilson and Kelling explained the central idea behind their theory in the following way: "At the community level, disorder and crime are usually inextricably linked, in a kind of developmental sequence. Social psychologists and police officers tend to agree that if a window in a building is broken *and is left unrepaired*, all the rest of the windows will soon be broken. This is as true in nice neighborhoods as in run-down ones. . . . One unrepaired broken window is a signal that no one cares, and so breaking more windows costs nothing."[1] Thus, the authors hypothesized, small signs of disorder may lead to more serious criminal offenses.

Empirical support for the "broken windows theory" had been provided many years earlier, in 1970, by Philip Zimbardo. Zimbardo arranged to have two automobiles without license plates parked with their hoods up, one in the Bronx and one in Palo Alto, California. The car in New York started to be stripped within ten minutes, and within three days it was totally destroyed. In contrast, the car in Palo Alto sat untouched for more than a week. Zimbardo then smashed its windshield with a sledgehammer. Within hours this car too was stripped, turned upside down, and virtually destroyed. Why the initial difference between the two locations? Zimbardo hypothesized that this was due to the social anonymity that could be taken for granted in New York. When such social anonymity is present, he argued, there is only a minimal need for "releaser cues" (such as the absence of a license plate and a raised hood) to ignite our darkest impulses. When such social anonymity cannot be taken for granted, it is necessary to provide more extreme "releaser cues" (such as smashing a window). Once the appropriate level of cueing has occurred for a particular social environment, the result will be virtually the same.

In Zimbardo's experiment, we see the broken window hypothesis literally confirmed. Given the appropriate signal that the car was left untended, there immediately followed a melee of criminal activity. Further empirical confirmation for the broken windows theory has more recently been provided by Wesley Skogan, who found a direct correlation between disorder and robbery, based on survey data in four major American cities.[2]

The results obtained in the larger "experiment" with the broken windows theory in New York City were no less stunning. From the time the policy was instituted in 1994 until it was unofficially discontinued in early 1999 (following the accidental police shooting of an unarmed man), overall crime dropped by half and there was a remarkable 67 percent drop in murder. After three years of slight annual decreases in the number of murders in New York City from 1991 to 1993 (4 percent, 7.4 percent, and 2.4 percent, respectively), from 1994 to 1998 there were double-digit annual decreases during the entire time that the broken windows policy was in effect (see table 1). By 1998, New York City boasted its lowest annual murder rate since 1966.

Of course, it is true that the national murder rate also began to fall in 1994, fueled in part by the fact that New York City's large population gave it an outsized effect on national crime statistics—but also by the fact that many other large American cities also experienced a decrease in murder during the 1990s. Nonetheless, the numbers put up by New York City are impressive in that they represent the largest decrease in total crime of any large American city since 1993. Indeed, from 1993 to 1998 the rate of decrease in the number of murders in New York City was more than twice that of the nation as a whole.[3]

Table 1

Number of Murders 1990–2000

Date	National	Change	New York	Change
1990	23,440	—	2,245	—
1991	24,700	+5.4	2,154	–4
1992	23,760	–3.8	1,995	–7.4
1993	24,530	+3.2	1,946	–2.4
1994	23,330	–4.9	1,561	–19.8
1995	21,610	–7.4	1,177	–24.6
1996	19,650	–9.0	983	–16.5
1997	18,210	–7.3	769	–21.8
1998	16,974	–6.8	633	–17.7
1999	15,522	–8.6	671	+6.0
2000	15,517	No change	673	No change

Source: Federal Bureau of Investigation, *Crime in the United States* Annual.

To what might one attribute such amazing success? One direct effect of the application of the broken windows theory in New York City was that as arrests increased for the petty offenses that had previously been tolerated, it was found that many of the arrestees were wanted on more serious criminal charges or were illegally carrying firearms. Another less direct but more salient linkage was appreciated in 1997 with the arrest of a serial murderer whose only previous arrest (and the fingerprinting that ultimately led to him) had been for jumping over a subway turnstile three months earlier. It is also probably true that there were many others among those arrested for petty offenses, who—although not yet felons—later would have graduated to more serious criminal offenses had they not been stopped. Can one be certain that such salient factors represent a causal link between the implementation of the "Quality of Life" campaign and the reduction in the murder rate in New York City? There are several complicating factors that make this nearly impossible to prove.

For one thing, the results of New York City's experiment in reducing crime were very visible nationally. And once it was clearly a success, a team of New York consultants began working with police departments in other cities across the nation in an effort to reduce crime elsewhere. One such effort in New Orleans resulted in a 23 percent drop in violent crime in the year after the changes recommended by the New York team were instituted; a similar result was appreciated in Baltimore in 2000, resulting in a 14 percent drop in homicide in what had previously been America's most murderous city. Indeed, once the cat was out of the bag and the broken windows theory started to be implemented in many cities across the country, perhaps it was no longer meaningful to assess the theory's effectiveness by measuring the numbers put up by New York City against the national average, for after the first few years, the national numbers themselves may have been affected by wider implementation of the broken windows theory.[4]

Such factors complicate both the empirical assessment of the broken windows theory and the worthiness of its consequent emulation as a paradigm case of effective public policy based on empirical social science. Yet, obviously, the police officials who instituted the broken windows theory were interested in decreasing crime, not in running a controlled social experiment. Thus, the conclusions that we may draw from even such a seemingly clear-cut example of policy success are imperfect in their implications for the role played by empirical social science and warrant continued scrutiny.

Unfortunately, there probably exists no perfect example that we may use to assess the prospects for empirical social science in ameliorating our social problems. And yet the results just cited are dramatic and heartening, and justifiably suggest a reason for optimism about the future role that social science might play in shaping public policy. And even if we find the example unconvincing and remain skeptical that the broken windows theory could possibly be responsible for such a dramatic reduction in crime, shouldn't this fuel our imagination in trying to discover the actual causal roots of such a stunning and laudable social outcome? Indeed, what better results might await us with an even more carefully controlled scientific study? Will we be able to emulate such success in the other areas of our social misfortune? Are we ready to start on the road that may lead us to the better world that we have imagined?

In this book, I have tried to outline a path that may lead us from our current ineptitude to the possibility of more social successes, like the one just cited of crime reduction in New York City. I contend that we will only get there, however, by faithful employment of the scientific attitude in reforming the way that we study human behavior, and the resolve to apply our findings to the creation of a better society once we have them. But we must not tarry. The task that now faces us is an urgent one, for the world is burning while we look for answers to the questions that have been put by social science. Those who deny this are either ignorant or callous. The job ahead may seem daunting, and our prospects for success may at times look bleak. It is well to remember that it similarly looked bleak for the natural sciences in the depths of the Dark Ages. The discovery of truth is not inevitable. It will take our greatest efforts to face ourselves honestly and solve the social problems that today cause so much human suffering. Will we heed this call to arms? Will we liberate ourselves from these current Dark Ages? Will we at last choose to put the social sciences on their true empirical footing? The fate of human existence on this planet may well depend on how we answer such questions.

A science of human behavior can lead the way out of the current mess of unreason and tragedy that hangs over human affairs. The application of our highest form of reason, science, to the study of our social problems is our best hope for salvation. Even in a dark age, our reason can see us through. Our future may well be brighter than we have imagined it, for scientific inquiry is well equipped to answer the questions that have been put by human misery. The world awaits our response.

Notes

Chapter 2

1. One of the most eloquent accounts of how this might be done in natural scientific practice is that of Karl Popper in *Conjectures and Refutations: The Growth of Scientific Knowledge* (New York: Harper Torchbooks, 1963).

2. Despite the attractive features of Karl Popper's account of falsification in helping us to make sense of what might be distinctive about science, it is important to point out that there has been great controversy within the philosophy of science over the adequacy of this model. While it is widely agreed that science respects the power of negative evidence, and tests its theories accordingly, scholars remain divided over Popper's particular account of this as well as his claim that this is the distinguishing feature of scientific theories.

3. James Harvey Robinson, *The Mind in the Making: The Relation of Intelligence to Social Reform* (New York: Harper, 1921), p. xi.

4. Respectively, Michael Fix and Jeffrey Passel, *Immigration and Immigrants: Setting the Record Straight* (Washington, D.C.: Urban Institute, 1994); George Borjas, "Immigration and Welfare, 1970–1990," *Research in Labor Economics* 14 (1995): 253–289, and George Borjas and Stephen Trejo, "Immigrant Participation in the Welfare System," *Industrial and Labor Relations Review* 44 (1991): 195–211; and Julian Simon, *The Economic Consequences of Immigration* (Oxford: Basil Blackwell, 1989).

5. Charles Leslie, "Scientific Racism: Reflections on Peer Review, Science and Ideology," *Social Science and Medicine* 31 (1990): 891–912.

Chapter 3

1. My own reaction to this idea is perhaps here instructive, for I am a liberal Democrat, yet I have never found anything disturbing about the idea that there can be a science of human behavior. Indeed, my hope has always been that good empirical social science would provide the basis for more effective public policy, that is less subject to wishful thinking or the political whims of the moment, and therefore may serve the long-term interests of those who have been disadvantaged.

2. Jacob Bronowski, *The Identity of Man* (London: Heineman Educational Books, 1965), p. 6.

3. Steven Fraser (ed.), *The Bell Curve Wars: Race, Intelligence, and the Future of America* (New York: Basic Books, 1995), p. 1.

4. Two of the best are those written by Myron Hofer (a psychiatrist), "Behind the Curve," *New York Times*, December 26, 1994, and Thomas Sowell (an economist), "Ethnicity and IQ," in Fraser, *The Bell Curve Wars*, pp. 70–79.

5. Given the political reality in most universities today, however, the majority of potential bias among social scientific researchers probably comes from the left. In a 2005 study by political scientists Stanley Rothman, S. Robert Lichter, and Neil Nevitte, "Politics and Professional Advancement Among College Faculty," *The Forum* 3, (1) article 2 (using data from the North American Academic Survey), it was found that 72 percent of American college faculty identify themselves as liberals, rising to 87 percent for faculty at elite schools.

6. Of course, one potentially complicating factor here might be the belief that the death penalty is also needed because it satisfies our desire to punish those who have committed heinous crimes. Since this attitude would reinforce belief in the appropriateness of the death penalty, it might tend to inhibit one's willingness to concede that the death penalty does not promote general deterrence, since it is felt to be appropriate anyway.

7. A typical view was recently expressed by Governor Mitt Romney, who is trying to reinstate the death penalty in Massachusetts. When confronted by a journalist with the fact that there is little evidence to support the view that the death penalty deters crime, Romney insisted that this was wrong. "Studies can show whatever you want them to show. Punishment has an impact on action, and the idea that a more severe punishment would have an impact on action is obvious to even a schoolchild. There's absolutely no question but that the death penalty would reduce a certain number of heinous crimes." Karen Olsson, "Death Wish," *The Boston Globe Magazine*, January 1, 2006, p. 20. The best resource on the debate over the deterrent value of the death penalty is Hugo Bedau, *The Death Penalty in America* (New York: Oxford University Press, 1982).

8. The problem, unfortunately, is not just that we have not had much rigorous empirical social science to date, but also that much of the good work that is now available has been blithely ignored by those who might be in a position to use it. This, of course, is what resistance to knowledge is all about. To the extent that there have been scientific answers to empirical questions about human behavior, many of those who create public policy (and therefore are in a position to use this knowledge) have felt free to reject those findings that do not suit their political intuitions (sometimes, ironically, on the grounds that social scientific research is too politically biased to rely on).

9. Gary Kleck, *Point Blank: Guns and Violence in America* (New York: A. de Gruyter, 1991), pp. 106–107.

10. Ibid., p. 122.

11. Ibid.

12. Ibid., pp. 127–129.

13. Cf. especially pp. 537–544 of Don Kates et al., "Guns and Public Health: Epidemic of Violence or Pandemic of Propaganda?" *Tennessee Law Review*, 62 (1995): 513–596.

14. Though it should be pointed out, in fairness to Kleck, that Cook's estimate is based on a nonanonymous survey conducted by the Justice

Department, which put many of the respondents who admitted to a DGU in the uncomfortable position of confessing what may have been a criminal act to a law enforcement official.

Chapter 4

1. It was in this letter that Galileo outlined his famous doctrine, which came to be known as the Galilean principle, that science and religion should be separate, lest religious faith be questioned as a result of scientific advancement. This position was finally adopted by the Catholic church in 1890. Pope John Paul II officially apologized to Galileo in 1992, 350 years after his death.

2. Galileo's book remained on the Catholic church's *Index of Prohibited Books* until 1835.

3. An excellent and detailed account of Galileo's trial before the Roman Inquisition can be found in Stillman Drake, *Galileo* (New York: Oxford University Press, 1980).

4. Of course, Galileo's advocacy of the importance of evidence in testing a scientific theory predated any models in the philosophy of science. It would be misleading, therefore, to claim Galileo as a witness in the debate between falsificationism and its critics. Nonetheless, it is true to say that Galileo was a strong advocate of the power of evidence in differentiating between competing empirical hypotheses, however that might be understood.

5. It is important to keep in mind that in Galileo's time, religion *was* political. It should also be remembered that political (and religious) ideology continues to influence natural science even today.

6. The full text of Act 590 can be found in M. Ruse (ed.), *But Is It Science? The Philosophical Question in the Creation/Evolution Controversy* (Amherst, N.Y.: Prometheus Books, 1996), pp. 283–286.

7. Other than Ruse, one might also examine P. Kitcher, *Abusing Science: The Case against Creationism* (Cambridge, Mass.: MIT Press, 1982), and D. Futuyma, *Science on Trial: The Case for Evolution* (Sunderland, Mass.: Sinauer, 1982).

8. This quotation is taken from Judge Overton's decision in *McLean* v. *Arkansas*, the case in which the constitutionality of Act 590 was tested. This is reprinted in Ruse, *But Is It Science*, pp. 307–331. The quotation is on p. 320.

9. *Time*, August 30, 2004.

10. Scientism can be defined as the slavish imitation of the trappings of science, while ignoring its true spirit. Cf. F. A. Hayek, *The Counter-Revolution of Science: Studies on the Abuse of Reason* (Indianapolis: Liberty Press, 1979).

11. Indeed, this was the case in the transition from Ptolemaic to Copernican astronomy. See T. Kuhn, *The Copernican Revolution* (Cambridge, Mass.: Harvard University Press, 1957).

Chapter 5

1. James Q. Wilson and George L. Kelling, "Broken Windows," *Atlantic Monthly* (March 1982): 31.

2. Wesley Skogan, *Disorder and Decline: Crime and the Spiral of Decay in American Neighborhoods* (New York: Free Press, 1990).

3. From 1993 to 1998, the national homicide count fell by 31 percent. During the same period, the number of murders in New York City dropped by 67 percent (see Table 1).

4. Does the drop in the national murder rate since 1993 reflect the increasingly nationwide effect of the broken windows theory? Or does it instead suggest that the numbers in New York are a function of some as-yet-unexplained national trend? If the latter, one must account for the fact that New York's numbers led the nation in every year since the new policy went into effect.

Bibliography

Baird, Robert, and Stuart Rosenbaum (eds.). *Punishment and the Death Penalty: The Current Debate*. Amherst, N.Y.: Prometheus Books, 1995.

Becker, Gary. *The Economic Approach to Human Behavior.* Chicago: University of Chicago Press, 1976.

Bedau, Hugo (ed.). *The Death Penalty in America*. New York: Oxford University Press, 1982.

Borjas, George. "Immigration and Welfare, 1970–1990." *Research in Labor Economics* 14, (1995): 253–289.

Borjas, George, and Lynette Hilton. "Immigration and the Welfare State: Immigrant Participation in Means-Tested Entitlement Programs." *Quarterly Journal of Economics* (May 1996): 575–604.

Borjas, George, and Stephen Trejo. "Immigrant Participation in the Welfare System." *Industrial and Labor Relations Review* 44 (1991): 195–211.

Bronowski, Jacob. *The Identity of Man*. London: Heineman Educational Books, 1965.

Chang, Iris. *The Rape of Nanking: The Forgotten Holocaust of World War II*. New York: Basic Books, 1997.

Close, F. E. *Too Hot to Handle: The Race for Cold Fusion*. Princeton N.J.: Princeton University Press, 1991.

D'Souza, Dinesh. *Illiberal Education: The Politics of Race and Sex on Campus*. New York: Free Press, 1991.

Drake, Stillman. *Galileo*. New York: Oxford University Press, 1980.

Federal Bureau of Investigation. *Crime in the United States*. Washington, D.C.: U.S. Government Printing Office. Annual.

Finocchiaro, Maurice. *The Galileo Affair*. Berkeley: University of California Press, 1989.

Fix, Michael, and Jeffrey Passel. *Immigration and Immigrants: Setting the Record Straight*. Washington, D.C.: Urban Institute, 1994.

Fraser, Steven (ed.). *The Bell Curve Wars: Race, Intelligence, and the Future of America*. New York: Basic Books, 1995.

Futuyma, Douglas. *Science on Trial: The Case for Evolution*. Sunderland, Mass.: Sinauer, 1982.

Gaffron, Hans. *Resistance to Knowledge*. San Diego: Salk Institute for Biological Studies, 1970.

Gross, Paul, and Norman Levitt. *Higher Superstition: The Academic Left and Its Quarrels with Science*. Baltimore: Johns Hopkins University Press, 1994.

Gross, Paul, Norman Levitt, and Martin W. Lewis (eds.). *The Flight from Science and Reason*. New York: New York Academy of Sciences, 1996.

Harris, Sam. *The End of Faith: Religion, Terror, and the Future of Reason*. New York: Norton, 2004.

Hayek, F. A. *The Counter-Revolution of Science: Studies on the Abuse of Reason*. Indianapolis: Liberty Press, 1979.

Herrnstein, Richard, and Charles Murray. *The Bell Curve: Intelligence and Class Structure in American Life*. New York: Free Press, 1994.

Hogben, Lancelot. *Retreat from Reason*. New York: Random House, 1937.

Huizenga, John. *Cold Fusion: The Scientific Fiasco of the Century*. Rochester, N.Y.: University of Rochester Press, 1992.

Kahneman, Daniel, et al. *Judgment under Uncertainty: Heuristics and Biases*. Cambridge: Cambridge University Press, 1982.

Kates, Donald (ed.). *Firearms and Violence*. San Francisco: Pacific Institute for Public Policy, 1984.

Kates, Donald, et al. "Guns and Public Health: Epidemic of Violence or Pandemic of Propaganda?" *Tennessee Law Review* 62 (1995): 513–596.

Kelling, George, and Catherine Coles. *Fixing Broken Windows: Restoring Order and Reducing Crime in Our Communities*. New York: Free Press, 1996.

Kitcher, Philip. *Abusing Science: The Case against Creationism*. Cambridge, Mass.: MIT Press, 1982.

Kleck, Gary. *Point Blank: Guns and Violence in America*. New York: A. de Gruyter, 1991.

Kuhn, Thomas. *The Copernican Revolution*. Cambridge, Mass.: Harvard University Press, 1957.

Levitt, Steven, and Stephen Dubner. *Freakonomics: A Rogue Economist Explores the Hidden Side of Everything*. New York: HarperCollins, 2005.

Lindblom, Charles. *Usable Knowledge: Social Science and Social Problem Solving*. New Haven, Conn.: Yale University Press, 1979.

Lynd, Robert. *Knowledge for What? The Place of Social Science in American Culture*. Princeton, N.J.: Princeton University Press, 1939.

Mallove, Eugene. *Fire from Ice: Searching for the Truth behind the Cold Fusion Furor*. New York: Wiley, 1991.

McIntyre, Lee. *Laws and Explanation in the Social Sciences: The Case for a Science of Human Behavior*. Boulder, Colo.: Westview Press, 1996.

Passel, Jeffrey, and Rebecca Clark. "How Much Do Immigrants Really Cost? A Reappraisal of Huddle's 'The Cost of Immigrants.'" Working paper, Urban Institute, February 1994.

Peat, F. David. *Cold Fusion: The Making of a Scientific Controversy*. Chicago: Contemporary Books, 1989.

Pinker, Steven, *The Blank Slate: The Modern Denial of Human Nature.* New York: Viking, 2002.

Popper, Karl. *The Logic of Scientific Discovery.* New York: Harper Torchbooks, 1959.

Power, Samantha. *A Problem from Hell: America and the Age of Genocide.* New York: Harper Perennial, 2003.

Robinson, James. *The Mind in the Making: The Relation of Intelligence to Social Reform.* New York: Harper, 1921.

Ruse, Michael. *But Is It Science? The Philosophical Question in the Creation/Evolution Controversy.* Amherst, N.Y.: Prometheus Books, 1996.

Sagan, Carl. *The Demon Haunted World: Science as a Candle in the Dark.* New York: Random House, 1995.

Simon, Julian. *The Economic Consequences of Immigration.* Oxford: Basil Blackwell, 1989.

Skogan, Wesley. *Disorder and Decline: Crime and the Spiral of Decay in American Neighborhoods.* New York: Free Press, 1990.

Snow, C. P. *The Two Cultures.* Cambridge: Cambridge University Press, 1959.

Taubes, Gary. *Bad Science: The Short Life and Weird Times of Cold Fusion.* New York: Random House, 1993.

Wegner, Daniel. *The Illusion of Conscious Will.* Cambridge, Mass.: MIT Press, 2003.

Wilson, Edward. *On Human Nature.* Cambridge, Mass.: Harvard University Press, 1978.

Wilson, James. *Crime and Public Policy.* San Francisco: Institute for Contemporary Studies, 1983.

Wilson, James, and Richard Herrnstein (eds.). *Crime and Human Nature.* New York: Simon and Schuster, 1985.

Wilson, James, and George Kelling. "Broken Windows", *Atlantic Monthly* (March 1982): 29–38.

Wootton, Barbara. *Testament for Social Science: An Essay in the Application of Scientific Method to Human Problems*. New York: Norton, 1950.

Wright, Robert. *The Moral Animal: Evolutionary Psychology and Everyday Life*. New York: Vintage Books, 1994.

Zimbardo, Philip. "The Human Choice: Individuation, Reason, and Order versus Deindividuation, Impulse, and Chaos." In W. Arnold and D. Levine (eds.), *Nebraska Symposium on Motivation 1969*. Lincoln: University of Nebraska Press, 1970.

Index